THE BISSES OF VALAIS

Man-made watercourses in Switzerland

Guy Bratt

Published by the author

British Library Cataloguing in Publication Data.
A catalogue record for this book is available from
the British Library.

First published in Great Britain in 1995
ISBN 09524984 0 5

Copyright © Guy Bratt 1995

All rights reserved. No part of this publication may be reproduced, stored
in a retrieval system, or transmitted in any form or by any means,
electronic, mechanical, photocopy or any other information or retrieval
system, without permission from the publisher, Guy Bratt.

Published in Great Britain
by Guy Bratt of Gerrards Cross,
Buckinghamshire, SL9 8PR.

Printed and bound in Great Britain by
the Amadeus Press Ltd.
517 Leeds Road, Huddersfield,
West Yorkshire, HD2 1YJ

To the memory of all those Valaisans who, for nearly 1,000 years, have built and maintained these remarkable watercourses and frequently lost their lives while so doing.

THE CANTON OF VALAIS

Fig. 1 The Swiss Rhône Basin

Contents

Preface 7
Using this book 9
Introduction 11
Note on Illustrations 14

Chapter 1 - *Climate and Geography* 15
Chapter 2 - *Chronology of Bisse - building* 25
Chapter 3 - *Construction and Maintenance* 33
Chapter 4 - *Administration and Regulations* 59
Chapter 5 - *Disputes* 69
Chapter 6 - *Water - rights and Irrigation* 75
Chapter 7 - *Numbers, Lengths and Names of Bisses* 81
Chapter 8 - *Recent Developments and Future* 85
Chapter 9 - *21 Suggested Excursions* 99

Maps. 111
Appendix - List of Bisses 117

Glossary 129
Acknowledgements 137
Bibliography. 139

Index 141

REFACE

About thirty years ago, my wife and I first spent some time in Canton Valais - comprising the Rhone valley in Switzerland and its many side valleys – instead of rushing through it by train or car on the way to somewhere else. As a result of that trip we and our three children, all of us pretty determined mountain types in our various ways, go there whenever we can and regard it as a second home. Consulting 1:25,000 maps during that first visit revealed watercourses which followed contours in a thoroughly unnatural manner and this was my introduction to the bisses, the man-made watercourses which irrigate much of the Valais.

Theo Schnyder, a former cantonal engineer, gave the essentials of a bisse as being at least 1,000 metres long and providing a minimum of 15 litres a second from its source – a "prise d'eau" or "Schöpfi" in a torrent - to the piece of land to be irrigated. The word bisse (masculine) has belatedly crept into the most modern editions of Larousse: to the best of my knowledge, it is not used anywhere outside the Western, French-speaking, part of the Valais. Its origin is obscure. Larousse, having ignored it until recently, now relates it to an old word meaning a snake, while a former cantonal archivist at Sion, maintaining that in ancient documents it appears as bez or beiz, would corrupt the z into ds, write the resulting word beds, pronounce it bids, relate it to the German Bett and deduce therefrom that it meant the bed of a watercourse. I find this explanation contrived and improbable.

In part, but not all of the Eastern, Swiss-German-speaking half of the Canton, the equivalent, not to be found in my Muret-Sanders German dictionary, is suon(e) or suen(e), feminine, with plural suonen or suenen. The origin is equally obscure but there is an ancient Indo-European root, su, meaning water, leading, 4000 years ago, to Hittite assu, meaning good or goodness, and to Sanskrit and then Turkish su, again meaning water. However, suon/suen is not found in documents earlier than 1500. Elsewhere in the Eastern part (e.g. Mund) the words used are Wasserleite, Wasserfuhr or local variants of these. Other specialised words used in this book are listed in the Glossary. Neither the book nor the Glossary, however, contain all the relative terms of art. There are considerable variations in the vocabulary used from one bisse to the next. This reflects their individuality and the isolation of the communities for which they were built. In the French-speaking part of Valais, the degree to which the patois itself survives also varies, e.g. it is used much more in Evolène and Savièse than in Ayent. In the Swiss-German-speaking part there are, similarly, different local usages in the very live Walliserdütsch of the area.

Thousands of winter visitors to the Valais have never seen a bisse – or suon – because the bisses obviously get buried in snow; many summer visitors remain ignorant of them for all sorts of reasons, e.g. they do not walk far enough; they have not the countryman's eye for what is not "natural"; they have not had their attention called to bisses by tourist literature, which only relatively recently has seen fit to mention them; they are perhaps not naturally inquisitive. We were intrigued from the start and looked in bookshops for elucidation. All we could find was one novel, "An Heiligen Wassern" by J. C. Heer, published in Stuttgart and Berlin in 1912 and given to me by Lily Fischer, a Swiss friend in Luzern. We have walked many bisses, some of them many times, and it was our continuing ignorance which led the children, all by then adult, to indicate rather forcefully that here was something for me to set about in my retirement.

I first approached the Archaeological Museum in Sion, the capital of the Canton, where I was told that, sadly, they had "quasi-rien" on the subject, but they sent me to the Canton Library nearby as the most likely source. That library holds about 50 titles of books and articles about or touching on the subject. I have consulted all of them; a very few are learned scientific works which go into great detail of climate and geology; some are of a more general nature, useful but sometimes rather superficial and mostly somewhat dated; some are so rare and valuable that they cannot be borrowed but have to be looked at in the library's reading room; some deal with one or a few individual bisses. I also consulted the Schweizerische Landesbibliothek (Federal Library) in Bern, who very kindly sent me one photocopy and their list of titles and suggested, correctly as it turned out, that the Canton Library in Sion would be found to be a more fertile source.

To pressure from the family and friends was added encouragement from cooperative staff of the Sion Library, enthusiasm from Jean Travelletti, the Secretary of the Commune of Ayent, and Firmin Morard, the President of the Grand Bisse d'Ayent, and friendly help from officials of some other communes.

Hence what follows.

Using This Book

Those who visit Valais may either follow the order of the book and proceed logically from the background material on geography through history to the construction, maintenance and administration of the bisses, or follow my journey of discovery and start walking some of the major bisses listed in Chapter 9 and then examining that background.

Those who have no present intention of visiting Valais should read straight through - *and perhaps the book will make them change their minds....*

NTRODUCTION

God was walking in the Valais one hot, fine day and saw how parched and burnt the meadows and fields on the slopes were. To a sweat-soaked peasant who came the other way he said: "It is much too dry in your country; I must make it rain again here very soon". The peasant answered thoughtfully: "Oh, no, Lord, don't do that, for we understand irrigation better." This conceit vexed God so much that, from that moment on, he punished the Valais with a sky from which rain very rarely falls.

Many years later, when Ausserberg was suffering even more than usual from terrible drought, the good people heard of a magician in Sion, the Canton's capital, who had helped other communes in times of great trouble. At a referendum they decided to send three well-regarded members of the council of their commune to see him and ask for help. The three delegates took food and wine for the journey, found the magician in Sion and told him of the plight of their village. When he had listened to them sympathetically, he said: "You are brave, tough people up there on your rugged, dry mountain and you deserve to be helped." He went to another room and, after busying himself for some time with all sorts of secret things, came back with a little box and said in portentious tones: "Something is hidden in this little box. Do not open it until you reach the place where you would like there to be a plentiful spring." The men accepted the box with joy, paid the bill, politely took their leave and set forth for home. Without a word, they reached a point near Leuk and there ate the rest of their food and drank the rest of their wine. They looked at their box from all angles and gradually curiosity about its contents got the better of them. They agreed to open it just the tiniest bit and then shut it again, but alas, as soon as they opened it, a brightly-coloured butterfly flew out and, while they looked on open-mouthed, it settled on the earth a few metres away. There was a great underground rumbling and thundering and a few minutes later a beautiful spring burst out of the dry, stony ground. In frightful consternation the men went back to the magician in Sion, but he sent them away, saying: "You have had good luck once but will never have it again." So they returned miserably to Ausserberg and, having decided not to lie about what had happened, drank deeply in their cellars in search of courage and owned up. They were never re-elected to the council of the commune and, down below, the spring, called in the dialect "Üüsserbärgerro Brunno" (Ausserberger Brunnen) still flows happily but uselessly into the Rhone.

In about AD 1300 the inhabitants of Lens were faced with a most pressing need for an increased water supply. They came to an arrangement with the fairies that the latter would build them a watercourse through excessively difficult and dangerous terrain, on condition that the Lensards would not ring their great church bell when the water arrived in their village. However, when it did, they could not restrain their enthusiasm and rang the bell. The fairies, furious, destroyed the water-course at once; the remains of it have been known ever since as the "Bisse des Fées".

However, in the collection of 2,344 stories from the Swiss-German-speaking part of the Canton published by Joseph Guntern, only eight mention bisses; even fewer appear in other collections, from either part of the Canton. In view of the importance of bisses to the lives of the people and, indeed, total dependence on them in some places, I find this surprising. However, the explanation of the experts, which I unreservedly accept, is that those things which one knows well, uses or makes and which, therefore, figure in one's daily life, do not lend themselves to legend or mythology. One is too close to them. On the other hand, the bisse as a scene of encounters with the hereafter occurs a little more often, mainly because watering was often done at night and night-time is more conducive to the products of fear and imagination. Glimpses of the light of lanterns, held by distant, unseen figures, working during moonless nights in precipitous places would surely give most of us a jolt.

Bisses have figured in elaborations of historical events, such as the miraculous recovery from the plague of Marie Rosset (or, according to another source, Rocher). She was a rich 15th century heiress, whose cure was brought about by the shock of falling into the ice-cold bisse from the bier on which she was being taken to be buried. In gratitude for her survival she gave her entire fortune to finance the completion of the Bisse de Savièse across the precipice of Branlires (see pages 35 & 49).

The object of this book is to provide for English-speaking readers a general account of bisses, including the reasons for their existence, the chronology of their building, the methods of construction, their administration and regulations, disputes about them, water-rights and irrigation, numbers, lengths and names, recent developments, their future, some suggestions for walking them and a record of at least some of the terminology used before it disappears from everywhere except the reference works produced by etymologists.

The total number and total lengths of bisses past and present are complicated matters and I have given them a chapter to themselves (Chapter 7). Suffice it to say here that Emmanuel Reynard of Lausanne University made a most comprehensive study in 1993/94 and, basing himself on the 1:25,000 maps and various inventories, lists 376, with an approximate total length of 1,740 km.

Although many bisses are still in use as essentials to agriculture and viticulture, modern tunnels, pipes and pumps have caused many others to fall into total or partial disuse and the same fate will doubtless befall many more. Only in recent years have some of those concerned appreciated the heritage and touristic value of these remarkable achievements of their forebears and, here and there, taken steps to preserve what remains of some the more impressive examples. On the completion of the new water supply of Ausserberg in 1972, the Blumlisalp (Thun) Section of the Swiss Alpine Club, with its Ortsgruppe Ausserberg, took over from the commune and now maintain the old Niwärch suon as a memorial to the courage and tenacity of generations of Ausserbergers. In 1991, a spectacular cantilevered section of the Grand Bisse d'Ayent, replaced by a tunnel in 1831, was restored in its original form by the Consortage du Grand Bisse d'Ayent (figs *1, 2, 3,* 32 & 33).

With some important exceptions, bisses are only operated during times when the land has to be irrigated; these times vary from place to place, according to the differing micro-climates and also depend on whether a bisse serves land used for vineyards or fields, or both. At Salgesch, for example, the water from the Raspille goes, from 24 June to August, to the vines by day and to the fields by night.

Although I have tried to produce a generalised account of the subject, I am conscious that, inevitably, I have given more space to the areas of the Canton which I know best. I hope that no offence will be taken by anyone who knows and enjoys some bisse or locality which I have not mentioned.

NOTE ON ILLUSTRATIONS

The illustrations are a mixture of modern and ancient. All the modern photographs except two were taken by members of the family and the copyright on them is ours. The two exceptions are appropriately attributed.

The ancient photographs had to be used because their subjects either no longer exist or have changed beyond recognition. They have been attributed to the writers of the books and papers in which I found them: the most recent of these was dated 1948. Clearly these attributions are not always correct, because some also appear in earlier documents with no attribution at all. My printers have done wonders with the old material.

CHAPTER 1

Climate and Geography

The reason for the existence of the bisses in Canton Valais is that such water as there is is frequently in the wrong places – remote from the limited cultivable land. While this state of affairs is obviously by no means unique – other occurrences of it caused the building of the Q'ANAT (underground irrigation channels) and JUBE (surface channels) in ancient Persia, the FALAJ in Oman, the LEVADAS in Madeira and irrigation systems in Spain, the Andes, the Himalayas and Pamirs - some of the bisses in Valais are undoubtedly most remarkable examples of tenacity, courage and sheer bloody-minded determination in human achievement.

The reason for the inadequacy of water supply is the climate of the area and that, in turn, is in large part determined by its topography and geology. The Valais canton is 110 km long, North-East to South-West, and averages 50 km in width, rising to 68 km at its widest, between Monte Rosa and Wildstrubel. In this relatively small area, about 5220 km^2, which comprises the greater part of the Swiss Rhone valley and its side valleys, there is a surprising variety of micro-climates. Temperature difference at the same altitude and the same time of day may vary greatly between points separated by only short distances. The Swiss Rhone valley lies between the Bernese Alps to the North and the Valaisan Alps to the South, the two highest mountain chains in Europe, which rise to heights of between 3,500 m. and 4,000 m. The height of the land above sea-level ranges from 372 m. where the Swiss Rhone, at its Western end, flows into the Lake of Geneva, to 4,600 m at the top of Monte Rosa. The valley floor itself rises only 1428 m from its Western end to the break of slope below the snout of the Rhone glacier (1800 m), 170 km to the East; it is, therefore, not only the longest but also the deepest valley in Switzerland. The two mountain chains influence the climate in two ways.

First, they are the cause of minimal and uneven rainfall over much of the area. When moisture-laden South and South-West winds meet these chains, the air is forced to rise and thus to cool. The water vapour which it carries condenses to form cloud and then to fall as rain or snow on the windward side and on the summits, leaving little to fall on the North-facing slopes of the Valaisan Alps and over the Valley itself, at least until the air is again forced upwards over the Bernese Alps, thus condensing the remaining water vapour into more rain-bearing clouds. This is known as the "rain-shadow effect" (fig. 2). The Rhone valley is a classic example. Itself in the rain shadow, it receives little rain. Instead, water reaches it in the form of unpredictable and ill-disciplined torrents, which run off rapidly from the heights to the valley floor and give little benefit to the surrounding country-side. This factor is not so marked in the area between the Simplon and the Furka passes because the Valaisan Alps there are not so high and the moisture-bearing south winds can get through and do not entirely unload until they reach the frozen summits of the Aar massif in the Bernese Alps.

Secondly, the effect of sunshine is increased on the northern slopes of the Rhone valley by the fact that the sun's rays strike them more or less at right angles, rather than obliquely as on the southern side. Their heating effect is, therefore, greater (fig. 3). To this must be added two supplementary factors:

> a) because the air coming from the South and South-West has lost much of its moisture content over the Valaisan Alps, as described above, fewer clouds form over the Rhone valley itself and the lower parts of the South-facing slopes of the Bernese Alps which, therefore, enjoy long hours of sunshine;

> b) there is much exposed schistose rock, which reflects the sun's rays like a mirror.

The variations of climate conditions caused by the funnelling of wind along valleys and over passes, and with which mountain-dwellers are familiar, produce local exceptions to the overall patern described above.

The hottest region of Valais is the central area, between Sion and Turtmann. The precipitation there averages only 60 and 68 cms. respectively a year. Although above 500m (Sion 542m., Turtmann 622m.), their average summer temperatures are 18.2° C and 17.5° C respectively. Some additional precipitation and temperature data are

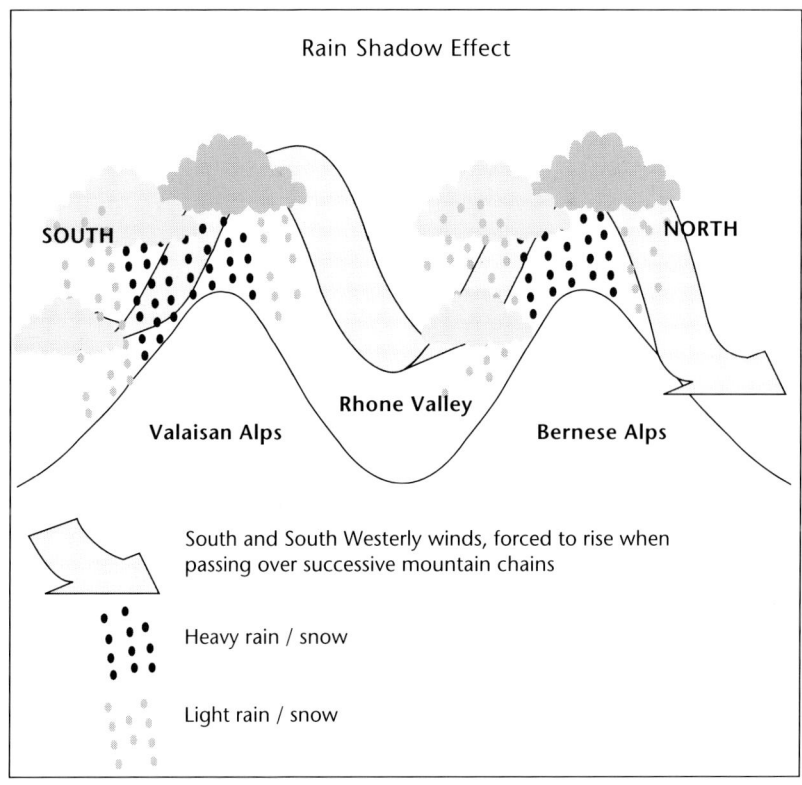

Fig. 2

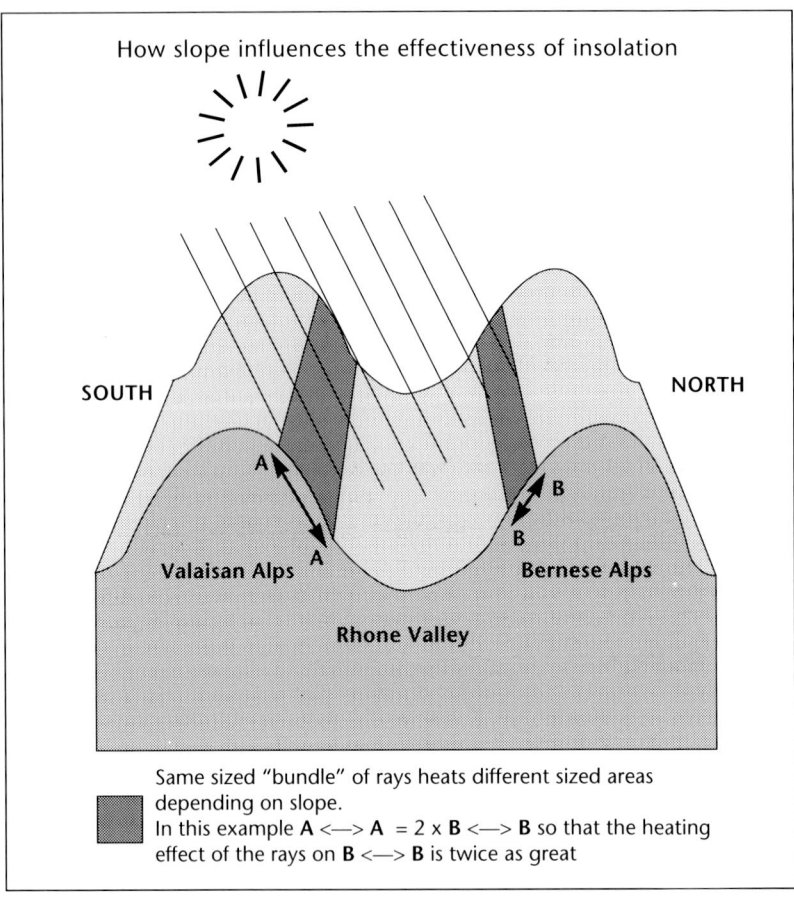

Fig. 3

given in fig. 4.

Before leaving the subject of climate, mention must be made of the foehn. Its squalls, coming from the south and having lost their humidity on the summits, just as the other moisture-laden winds referred to above, rush down into the valleys in violent gusts, so dry that they constitute a fire danger and so hot that they melt a lot of snow. There is a saying among some of the locals that one day of foehn is the equivalent in this respect of a fortnight of sunshine. Also, if the foehn blows at the appropriate time, the ripening of the grapes is appreciably accelerated.

The traveller by rail or road through the Rhone valley sees vast areas of fruit trees and vegetables and might well imagine that, with its rich alluvial soil, it had always been thus. This was far from being the case. The Rhone, in its natural state, was pushed from side to side in its valley by the sedimentary cones of its tributaries, changed its course unpredictably and flooded often and catastrophically, driving human habitation and such agriculture as there was up to higher ground - where no benefit could be drawn from the proximity of the river. There were 11 great floods between 1475 and 1902. The first serious attempt to control and canalise was brought about by a treaty between the Cantons of Valais and Vaud in 1836. The Federal Government first intervened with a subsidy in 1863 and thenceforth, with the help of high expenditure, the river has been disciplined and agriculture revolutionised. Certain areas, mostly of poor soil, have been left available as temporary receptacles for flood waters. The effectiveness of this policy was shown, for example, when the Saltina misbehaved in 1993: Brig was flooded disastrously, the flood plain filled, but little damage was done downstream.

All of which goes to show how exasperating it is that, with so much water about sometimes in the Valais, there is a shortage of it where it is needed. The peasant mentioned in the Introduction, whose conceit so offended the Almighty, has a lot to answer for.

The Alps are relatively young mountains; the gigantic upheavals which produced them shattered, up-ended and overlaid strata chaotically, so that what is now top surface presents a great variety of rock formations (very roughly 75% granite and crystalline schists, 20% sedimentary rocks and 5% alluvial), with an infinite range of permeability.

AVERAGE PRECIPITATION AND TEMPERATURE AT 5 METEOROLOGICAL STATIONS IN VALAIS

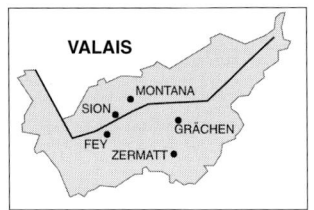

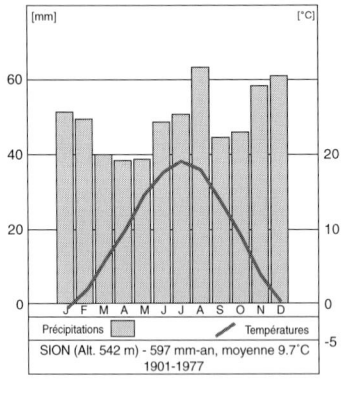

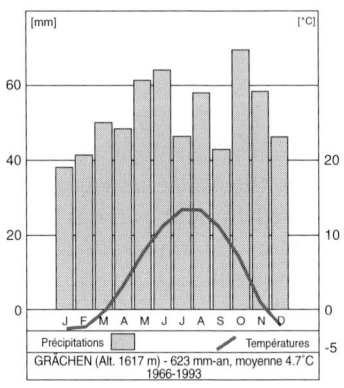

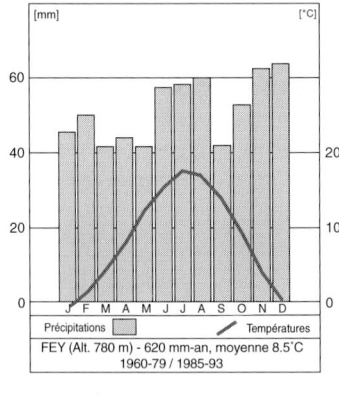

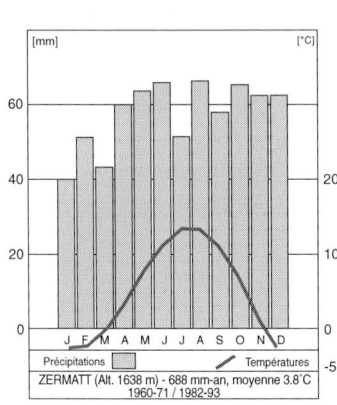

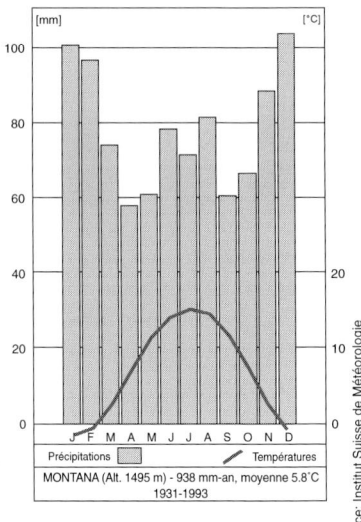

© Emmanuel Reynard, Institut de Géographie de l'Université de Lausanne, décembre 1994.

Fig. 4

The Swiss Rhone has nearly 200 tributaries. The lie of the land on the two sides of its valley is strikingly different. On the left bank, the first ten tributaries below the glacier are steep torrents but thereafter there are longer and, in some cases, quite wide side-valleys, running roughly South to North. Each contains a main, usually glacier-fed tributary, with many streams flowing into it, and most of these valleys rise relatively gently towards their heads in the high mountains. In some cases, however, there are marked "steps", e.g. the valley of the Trient ends in a spectacular water-fall down into the Rhone. These valleys were the earliest of the Rhone's side-valleys to be inhabited and cultivated. Some have at their southern ends passes over the watershed which were more readily passable in the Middle Ages than they now are; for instance the Col d'Hérens, which carried a good track from Zermatt to Evolène, the Col de Fenêtre, which carried a vehicle road linking the Vallée de Bagnes with Aosta and the Col de Collon which was used from as early as 1350 for driving cattle from Evolène to the Valpelline, on the way to fairs at Aosta. Some of the longest bisses were built on this side of the Rhone, e.g. those of Saxon (at 32 km, the longest of all), Le Levron (13 km), Chervé (or Chervaix, 15 km), Hérémence (10 km). Nendaz (19.5 km), Mayens de Sion (or Vex, 12 km), Heidenwasserleitung (or Païens, 15 km), which for many years from 1305 brought water from the Gamsa to the highest vineyards in Europe at Visperterminen), Augstbord (12 km), Niwa (Emdbach – Zeneggen 20km).

In contrast, on the right bank of the Rhone, the tributaries are torrents in steep gorges; some of these are glacier-fed, with relatively consistent flow, e.g. the Morge, while others depend largely on snow run-off and are, therefore, much less regular, e.g. the Sionne. The principal bisses on this side are La Tsandra (11.7 km), Savièse (11 km), Bitailla (or Bisse Taillaz, 4 km), Grand Bisse d'Ayent (15 km), Sion (13.5 km), Lentine et Mont d'Orge (7.5 km), Clavoz (7.7 km), Grand Bisse de Lens or Bisse de la Rioutaz (13.8 km), Ro (9.5 km), St. Léonin (7.5 km), Varen (7 km), the Ausserberg group (of which Niwärch, the longest, is 8.5 km), the Mund group (of which Badneri, the longest, is 6.4 km), Belalp-Nessel, Martisberg, and Lax. Some of these, notably La Tsandra, Sion, Lentine et Mont d'Orge, Clavoz, Varen, St, Léonin, serve almost exclusively vineyards rather than the more general agricultural purposes of the rest.

The contrast between the two sides of the Rhone valley is repeated in the few substantial secondary valleys which run very approximately parallel to it: Vallée de Bagnes in the Lower Valais and those of Lötschen and Binna in the Upper Valais. At Martigny, where the Rhone makes a 90 turn to the right, there is a very definite climatic change; the Bisse de Trient, above Martigny, is the most westerly one of all.

The chemical content of the suspended solids, averaging about 1 kilogram per cubic metre, in the water in bisses naturally varies with their origins. The water-borne mineral matter coming from crystalline rock is superior as a nutrient to that from limestone. Water emerging from glaciers ("Gletschermilch") is particularly rich in fertilising properties and the water from the southern valleys, especially of Bagnes and Turtmann,which passes through granite and schistose territory is so beneficial that for many years it was used instead of any other fertiliser. On the other hand, the low temperature of glacier water is a disadvantage if it is to be used for watering cattle on alpages because very cold water upsets their stomachs. Some warming-up occurs during the courses of the longer bisses and can also be achieved at watering points, which consist of several troughs, each one leading into the next, thus presenting a substantial surface area to sunshine and warm air. Good examples of this are in the alpages of Sérin near Les Rousses above Anzère, to which water is now brought from the Wildhorn glacier, nearly all the way in iron pipes, and fed in each case into rows of four or five connected troughs.

For many years up to the end of the early period of bisse building, i.e. about 1550, there had been – even by Valais standards – very low annual precipitation. Glaciers were retreating and many old-established pass routes were ice-free.

Then, from 1590/1600 until 1855/60 there occurred the "Little Ice Age". When the supply of ice from the firn area, where a glacier is nourished by snow and firn accumulation, exceeds the loss in the ablation area, where it is depleted, mainly by surface melting, a glacier advances. Such advances during the Little Ice Age obviously destroyed everything in their paths, including the prises d'eau and parts of the courses of some earlier bisses; e.g. the Grosser Aletsch glacier, which reached its maximum length in 1849,did that to the Oberriederin above Ried.

Although the relationship between changes of climate and fluctuations of glaciers is highly complex and their effects vary greatly from one glacier to another, it seems reasonable to assume that advances result in more water becoming generally available. Furthermore, in Valais, some cultivable lands and woodlands were lost to glaciers. These two factors reduced the need for bisses, so that fewer were built after 1600 and many more no longer properly maintained. For example, at the begining of the 19th century the Grand Bisse d'Ayent fell into such disrepair that it stopped flowing for seven years until the good Curé Cordel stimulated his parishioners to repair it. Similarly the Bisse du Levron fell into disrepair and had to be substantially rebuilt in 1834-38 and part of it again in 1860 and 1889. At the other extreme, retreating glaciers put some old bisses out of action because torrents took new courses inaccessible from the original prises d'eau. In passing, it is worth noting that, although most of the Alpine glaciers are now reduced to their minimum extent of the last 1,000 years, their behaviour in periods of advance and retreat is by no means consistent: the Rhone Glacier has never advanced during or since the Little Ice Age except for the period 1912-1922; the Giétro Glacier has never stopped advancing moderately until quite recently and in 580, 1549, 1595, 1640, and 1818 caused catastrophic floods with heavy loss of life. In 1977 of the 108 Swiss glaciers for which details were recorded, 52 were advancing, 42 were retreating and 14 were stationary; in 1988 of 107, the equivalent figures were 27, 74 and 6; in 1989 of 105, the figures were 19, 83 and 3.

Such, then is the natural background to the lives of those who built and used the bisses and depended on them for their livelihood.

Chapter 2

Chronology of Bisse-Building

After Julius Caesar's legions over-ran the Rhone valley, the Romans divided the area around the hills of Valère and Tourbillon between themselves and the people whom they found there, a Celtic tribe called Seduni. With the disintegration of the Roman Empire, chaos, disorder and destruction reigned in the Valais, like everywhere else; the next historically decisive event was the move by the Bishop of the Valais of his episcopal seat from Octodurum (Martigny) to Sion in AD 580. The reason for this move was that a glacial lake, which had been formed behind an ice and moraine barrier at a narrow point near Mauvoisin in the Drance valley, broke through that barrier and caused frightful devastation right down to Martigny.

In A.D. 999 King Rodolphe III of Burgundy gave, totally and inalienably, the county of Valais to Bishop Hugues of Sion, in exchange for feudal obligations. The Bishop thereby became a Prince and the temporal ruler of Valais from the source of the Rhone to Martigny. This gift, which was by no means unique, was to have serious consequences for the history of the canton; without it the independence movement, the conflict with Savoy and the structure of the "State of the VII dizains" would never have taken shape. Sometime between the middles of the 11th and 12th centuries, a monk named Ruodpert wrote "The Life of St. Théodule, Bishop of Sion". In that work was the first mention of an entirely legendary gift by Charlemagne of the county of Valais to St. Théodule; this gift, which never happened, was subsequently quoted under the name "La Caroline" whenever the church in Sion wanted to assert its rights over the canton. The first communal charter of the city, setting out the respective rights of the Bishops and the citizens, was established in about 1217. In an agreement of about 1293, between Count Amédée V of Savoy and the Bishop of Sion, "La Caroline" was mentioned for the first time as such and became a recurring source of discord, especially from the end of the 15th century, with the church striving to retain temporal power and the "francs patriotes" equally determined that it should not. Throughout much of this period there were frequent conflicts between the Bishop and the Dukes of Savoy and Sion was pillaged and plundered several times. Eventually the troops of the Swiss Confederation defeated the Savoyards at the Battle of La Planta in 1475 and Sion recovered the Lower Valais; thus the political unity of the Canton was achieved. The temporal power of the Bishop began to wane at the start of the 17th century and in 1634 Bishop Hildebrand II Jost was obliged formally to abandon any claim to such power in a document known as the Rénonciation à la Caroline.

In his temporal capacity the Bishop and his subordinates figure in the granting of water rights and the authority to build bisses. For example, in 1448 Prior Jean de Lens was requested by the Four Quarters of the Commune of Lens to build the Grand Bisse de Lens (or Bisse de la Rioutaz). The influence of the church is mentioned in more detail in Chapter 5.

Given the presence of the Romans in the Rhone valley, it is not surprising that claims have been made that they were the first builders of bisses but I have found no documentation to substantiate this. The analogy of bisses with Roman aqueducts elsewhere seems to me false because the former were all locally conceived and built, whereas the latter were the products of a highly centralised regime. Fritz Rauchenstein, an eminent builder and historian of bisses, mentioned the date 440 in connection with the Bisse d'Hérémence, but the consensus of opinion now seems to be that, in the context in question, a 1 in front of the figure was omitted in the printing and he meant 1440.

Similarly, the heathen (païens), Saracens, Huns and others not specified have been credited with the introduction of bisses, but all this seems to me to be a misconceived attempt to find a specific origin for a simple essential. Man cannot live without water, so, if there is some within reach – even if it has to be a long reach – he will naturally go and fetch it, especially if, by so doing, he can live and produce sustenance in an otherwise desirable place; in the Dark Ages some places were desirable simply because difficult of access by potential enemies. These attributions of origin all date from the second half of the nineteenth century; it seems to me more likely that water channels have existed in some form since the first human settlement in the area. Be all that as it may, bisses were certainly in use by the beginning of the eleventh century.

The Ernen chronicler Moritz Michel describes a lawsuit about water rights between the villages of Bellwald and Fürgangen in about 1008; twelfth century records mention grants of water-rights in the Val d'Hérens and two bisses on the left bank of the Lienne (identified later as those of Ro and Sillonin); thirteenth century records mention bisses at Ausserberg, Fiesch, Ernen, Milibach and Emd; in 1300 the Bisse de Ro was built to Lens, Bitailla (or Bisse Taillaz) – so-called because at its origin "taillé" (cut into) the rock – was referred to as already existing on the left bank of the Sionne and information is recorded at Salgesch about how the rights of neighbouring communes, including Varen, to the waters of the Raspille were defined by the Bishop of Sion. In 1307 two more bisses taking water from the Sionne, those of Lentine and Arbaz, and two bisses above Vercorin are mentioned as already existing. Another document from Ausserberg dated 1311 tells of the water-course known as "Chänilwasser" from the Bietschtal to Ausserberg, on which an accident to a maintenance party killed 12 men on one day and led to its replacement in 1361 by a new 8.5 km. channel – Niwärch (Neuwerk) and still so-called – from the Baltschiedertal. A chronicle of 1336 tells of a new aqueduct in Fiesch; the oldest document from Lax tells of one there in 1347. A 17-kilometre-long suon, "D'Bärgerwissa", drawing water from the Wanni glacier, already existed in 1351 when an agreement was made at the church of St. John the Baptist at Fiesch between the communes of Fiesch, Fieschertal and Martisberg for the building of a new suon mainly for the benefit of the last-named. On the opposite side of the Rhone valley at the northern end of the Vispertal, the Augstborderin or "Niwe" suon already existed in the fourteenth century, having replaced an earlier and even longer watercourse. Water-rights are mentioned in the will of Bishop Guichart Tavelli drawn up at Château de la Soie in 1366. In records of 1367 the Bisse de Saint-Léonin, above Saint-Léonard, is mentioned as already existing. In about 1385 the people of Chalais and Réchy built a bisse to take water from the Réchy torrent to their fields and in about 1390 the inhabitants of Vercorin did the same and this bisse fed three others which irrigated the fields and mayens of Grône and Loye.

In 1400 the Bisse de la Tsandra was built. Its prise d'eau is at 1,430 m. on the glacier-fed Morge, not very far from the point where a channel (sometimes also confusingly called the Bisse de La Tsandra) was cut through the rock in 1885 to reinforce the Netage, the not-always reliable source of the Bisse de Savièse on the opposite side of the gorge, and thus double the flow of that bisse. The courses of these two bisses, of approximately the same length (Savièse 11 km., Tsandra, 11.7 km.) could not be more different. La Tsandra throughout its length is an agreeable walk (though the bisse itself is now in a pipe) while much of Savièse is – or was until the 4.8 km. Prabé tunnel was opened in 1935 – a very difficult and unfriendly monster, to which I have unavoidably made many references, especially in Chapter 3, page 47. The Bisse de La Tsandra has two "tributaries": the Petit Torrent, which is diverted into it near My, and the Torrent de Trente Pas; the latter dries up each year towards the end of June. La Tsandra's water was always very precisely controlled by its "procureurs", who represented the sections of Daillon-Mouzerin, Daillon-Tsandraz, Premploz, Erde and Aven – villages which, without the bisse, could not have existed as it was their only source of water. Its modern history is unusual and it seems more appropriate to place it in Chapter 8 and you will find it on page 95. At an unknown date during the first half of the 15th century, the people of Lens built the Bisse du Ro (Rho, Rot, Roh, Ro – c.f. ru in Val D'Aosta patois) to bring water from the Ertense torrent at 1,750 m. to Lens. This bisse had a disturbed history. Costly and unsuccessful attempts were made in 1858, 1871 and 1925 to increase its flow by adding water from the Huiton ponds at 2,500 and 2,400 m. via a 300-metre tunnel to the Ertense and, in 1944, by boring a 2,000-metre tunnel through the Mont-la-Chaux but none of this was satisfactory, springs failed and both bisse and tunnel fell into disrepair. From 1949 the tunnel was used as a short-cut to and from an alpage and the walk-way along the bisse as a footpath, which became increasingly dangerous until thoroughly refurbished in 1962 as a tourist attraction. In 1430 the building of Savièse's Torrent Neuf (I Torin Nou, Bisse de Savièse, Bisse de Ste Marguerite) began under the direction of one Arnold von Leiggern (or Leukron), whose father had previously built a suon at Ausserberg. The original prise d'eau was at 1,660 m. on the Glacier-fed Netage, but the top section of the bisse proved unsatisfactory and the prise d'eau was later re-sited 300 m. lower down. The bisse also collected the flow from the Vauclusian Fontana Dzemma (twin springs). After 20 years work, the Torrent Neuf took the place of the much older "Croué Torin" (evil torrent) which was then abandoned and, surprisingly, some traces of it still remain.

In 1443 the people of Le Levron applied to the Abbot of Saint-Maurice for authority to build a new "trajou" – so there must have been one there earlier – from a source at Chardonnay but work did not begin

Fig 5: Bisse de Clavoz: in the vineyards near Champlan, above Sion. Note the high masonry wall supporting the upper vineyard.

until nearly 20 years later. In 1448 the archives of Lens tell of how the Prior, Jean de Lens, was requested by the four Quarters of the Commune de Lens to build the Grand Bisse de Lens (or Bisse de la Rioutaz) and how he did it in only two years, 1448 to 1450. In 1452 it was recorded that the Grand Bisse d'Ayent had recently been inaugurated (noviter inceptum) and the next year the bishopric and town of Sion cooperated to build the Bisse de Clavoz.(Figs. 5 & 6). Between 1480 and 1484 the Bisse de Riccard or Chararogne was built to serve Chalais and Réchy.

After this lively burst of bisse building in the Middle Ages, the process slowed down. Little new building (as distinct from maintenance) was done in the sixteenth century and practically none in the seventeenth and eighteenth. The reason was, it seems to me, related to a climatic variation (see comments on this in Chapter 1). Then, in the nineteenth century and on into the early years of the twentieth, construction and improvement revived, with improved building techniques and changing economic conditions. The Bisse de Saxon, at 32 km., the longest of all, was built between 1865 and 1889 to take water from the Printse to Saxon. It crossed the mayens of Nendaz, Isérables and Riddes without contributing to their irrigation. Another example of bisse building at this time is the Bisse-Syphon de Montorge at Sion, which

Fig 6: Bisse de Clavoz: Stream crossing the bisse in the vineyard near Molignon.

was inaugurated on 23 April 1895 and enables vines to be grown in an area which had everything that they needed except water. In 1901 the Leukerin suon was built to take water 4.3 km. to Leuk; in 1902 the Rischeri suon from Steinenbach to Ried-Brig; in 1903 the Bisse d'Orméan 1.5 km. from the Rèche to Réchy.

In that year also the last major bisse to be built, the Bisse de Sion, was constructed to add water to the flow of the Sionne, to make good the quantities taken from the springs known as "La Fille" and "La Fillette" and to increase the amount available for the orchards and vineyards of Sion and Savièse from the Bisse de Lentine (Figs **4 & 5**). The Bisse de Sion takes its water from 1,820 metres up and runs its full length of 13.5 kms on the territory of the commune of Ayent, which had no claim on its water. Two-thirds of its length required new work and the rest made use of the former Bisse des Audannes (or Ohannes), which was built by the communes of Sion and Ayent between 1859 and 1862, never functioned properly because of seepage and evaporation and was abandoned amid great embarrassment in about 1870. Unsurprisingly, it was and still is referred to as the "Bisse sec". (Not to be confused with another bisse sec – built in 1911-1912 by the people of Arbaz, to take the water from la Combaz to their fields – a project which caused such uproar in Ayent that, although completed, it had to be abandoned.)

In the very earliest days of bisse building, the question of costs did not arise; those in whose interests it was to build a bisse got together and did it themselves. Sometimes feudal overlords ordered the building of bisses, no doubt mainly in their own interests but perhaps, now and again, in the interests of their vassals. Sometimes those vassals petitioned to take over existing bisses. However, it was not long before materials had to be bought and paid hands, feudal superiors and pioneering entrepreneurs had to be remunerated. Amends had to be made to others whose property was damaged by leaking or overflowing bisses – a cause of perpetual disputes, some examples of which are given in Chapter 5. Payments were frequently made partly in cash and partly in produce of the land and, as the consortages and communes developed, very detailed accounts were kept by those responsible and many survive in the archives. However, except for the specialists in this field, the figures are almost meaningless because a plethora of different coinages was used through the centuries and their values varied enormously according to date and location. In Valais, different systems existed upstream and downstream of the Morge. Some of the moneys quoted were never struck but were simply units of account. Clément Bérard, in the last pages of his "Bataille pour l'Eau", sets out part of this jungle in some detail and then comments: "Allez donc vous y retrouver avec tout cela" ("Go and sort yourself out with that lot "). Suffice it to say that the cost of building, repairing and generally maintaining the bisses took a very substantial slice out of the resources of their users. Sometimes, this was beyond the means of those concerned, which led to the abandonment of some bisses but, from 1900 onwards the Canton and Federal Governments contributed large subsidies; for example, improvement to the 12km. Augstborderin in 1949 cost Fr.1,850,000, of which the Federation paid 50% and the Canton 35%. The works made possible by the subsidies have resulted in spectacular improvements in water supply; they have also saved many lives. Many tunnels have been bored to by-pass some of the more horrendous stretches of bisses which previously claimed victims year after year.

Further details of the bisses mentioned in this chapter and of all others still existing and walkable in whole or in part are given in the appendix.

CHAPTER 3

Construction and Maintenance

Obviously the first concern of those proposing to build a bisse was to find a reliable source of sufficient water. In relatively few cases water was taken from close to the snouts of glaciers, at above 2000 metres, but usually it was taken from a non-seasonal mountain stream ("torrent" or "Bach") at between 1,000 and 2,000 metres. Occasionally there was the luxury of a choice of source, the "goodness" of the water depending on what was carried in suspension in it and that, in turn, depending on the terrain through which it had flowed. Crystalline rock produces a more nutritious solution than limestone (it also helps to make bisses water-tight, both by depositing a partial sealant on the bottom and by filling interstices at the sides); glacier water ("Gletschermilch") is better than that from streams emerging from the ground – so much so that it was sometimes considered a fertilizer in itself. By no means all available water is suitable for drinking.

The second step was to select a place where the torrent was accessible for the deflection of water into the future channel (the tête du bisse, prise d'eau, leviour, Wasserfassung or Schöpfi), preferably via a sand-trap where the heaviest, useless solid matter could settle (figs 7 and *4*).

The third was to find a route for the bisse all the way to the place where the water was required. It was obviously desirable to keep the course of the bisse as high as possible for as far as possible, because this enabled water to be taken for the longest practicable distance and thus to irrigate the maximum area of land. In the case of the Bisse de Saxon, the longest, this was 32 km away. Especially but not exclusively in the initial parts of the course, obstacles were legion: moraines, rocks, landslides, scree slopes, alluvial cones, bogs, gorges and gullies, other streams which had to be bridged over or under because their water belonged to somebody else, and vertical or overhanging cliffs which had to be traversed. When the builders were faced with very difficult and/or dangerous sections of route which simply could not be avoided,

33

Fig 7: Bisse Taillaz (Bitailla): prise d'eau and distributor on the Sionne at Dorbon. The plaque on the right carries the arms of Ayent, Arbaz and Sion and the words "Prise et répartiteur du Bisse Taillaz 1985-1986".

these were and still are called "passages obligatoires". Another hazard which could affect routing was "le pas de la matta" (the girlfriend's step). If the engineers' sweetheart had some land which required irrigation, he would do his best to route the bisse above it if she was entirely cooperative, but below it if she was not. In one story, the girl's obstinacy led to the bisse ending in a waterfall and the girl remaining a virgin. In another more picturesque version, in which the "pas" becomes a "saut" (jump) the bise under construction is identified as the Bisse d'Ayent (1442) and the location as Pra Combère, the resultant waterfall was said to be 700 ft. high, but I have never been able to work out just where, in relation to Pra Combère, it was supposed to be. In this version, also, after all the engineer's attentions and eventual departure, the rich and beautiful heiress went on wearing a white ribbon in her best hat when she went to Mass, but had the grace to blush as red as a poppy. In 1484 the consorts of Le Levron made the egregious error of trying to build their bisse from its lower end and landed themselves with a quite unnecessarily difficult route.

Even where the course was to pass through relatively easy areas of meadow and woodland, the maintenance of levels – that is to say regular down-gradients of 2¹/₂ to 5% – without the benefit of surveying instruments presented difficulties. A lower figure could lead to rapid sanding-up of the channel; a higher one to uneven flow and flooding further along it. In areas such as the run of the Bisse de Savièse, with a steady gradient maintained throughout, across the vertical rock face of the Paroi des Branlires and the Paroi du Sapin, the achievement of this requirement was a most extraordinary feat. The primitive methods which were so effectively used for the medieval bisses included: rolling a pebble or marble along a plank; lying one one's back and adjusting the level to what one could see over one's own head; having a companion install himself, sometimes hanging on a rope, on the opposite side of a gorge shouting out the levels; using (e.g. for the sixteenth century Ebibergeri suon at Stalden) a simple instrument consisting of a triangle on which was mounted a plumb-line. Sometimes mistakes were made and much time and effort and some lives lost in vain. The course of some bisses is idyllic from start to finish; that of some other others is vertiginous and downright dangerous throughout; that of yet others an alarming mixture. When you walk through the forest up the course of the Bisse de Savièse, you pass the chapel of Ste. Marguerite at Barma de la Dzour, turn a corner a few metres beyond it and suddenly come to the edge of the Paroi des Branlires, referred to above, where the wonderful cantilevered wooden channel (figs 8 and 9), replaced in 1935 by the Prabé tunnel, 4,800 metres long, has totally disappeared after 500 years of use, leaving nothing but the abyss.

Fig 8: Bisse de Savièse. The priest accompanies a working party on a vertical rock wall (Paroi des Branlires) high above the Morge river.
(Photo: A. Vautrier)

Fig 9: Bisse de Savièse. The cantilevered crossing of the Paroi des Branlires. Some of the co-outacoués have been replaced by iron rods because, at the level of the channel, the rock was not solid enough to support the extra weight of the larger wooden co-outacoués which would otherwise have been necessary.
(Photo: Ch. Paris)

Fig 10: Blatten. Small-capacity channel (hollowed-out tree-trunk) carried on masonry columns. (Photo: A. Vautrier)

Naturally there was nothing like a "code of practice" for building bisses. Each was tackled ad hoc by those concerned (whether commune or consortage – see Chapter 4) and its problems addressed with little, if any, previous experience on the part of the builders. One notable exception to this was, again, the Bisse de Savièse, the construction of which, as mentioned in Chapter 2, was put in the hands of an "expert" from elsewhere in the person of Arnold von Leiggern who may be presumed to have learnt a lot from his father, who had previously built one of the suonen at Ausserberg.

Digging a channel through woodland and meadows presented few problems other than, first, coming upon and having to break and move or get round unforeseen boulders encountered just below the surface of the ground and, secondly, as mentioned above, maintaining levels. Material excavated, both during construction and afterwards, during the essential regular clearing of the channel, was thrown on to the banks, thus providing the walk-way (banquette or Tretschbord) for the gardien and continually reinforcing the "downhill" bank of the bisse, which was further stabilised by the root growth resulting from the increasing amount of well-irrigated soil. The crossing of shallow valleys and hollows was manageable either by mounting wooden troughs on simple supports or trestles (fig. 10 and 11) or by building masonry aqueducts

Fig 11: Niwärch. Channel carried on masonry columns and trestles. (Photo: F. Schmid)

such as those on the bisses de Clavoz and Lentine (fig. 5) and on the Bärgerwissa of Lax and Martisberg in Goms. If errors were made in such relatively friendly terrain, they could be corrected without much trouble but it was imperative to avoid any in the traverses of rock faces, where cantilevered wooden aqueducts had to be laboriously constructed.

To build these, two systems were used. The first was to select and hollow out trunks of larch or fir trees – preferably larch because their wood is more solid and less permeable – and cut their ends to fit each into the next. To support these trunks (bazots – or brozets, brochets, chénaux, tsénas, Kennel or Chänile, such are the local variations) 20 cm-square holes (boutzesses) were made in the rock face, to a depth of 15 to 20 cms, into which square beams (boutzets, butsets) were inserted and made fast with larch wedges. To these beams, selected naturally-hook-shaped pieces of wood (Krapfen or Chrapfen) were pegged and the hollowed-out tree trunks laid in them (fig. 12). The walk-way for the gardien or Wasservogt (Hüter) took the form of planks laid on top of the tree-trunks and sometimes tied to them with creepers and twisted branches (figs. 12 and 13). The second system, which could carry a much greater volume of water, consisted of open-topped channels of rectangular section made of larch planks. This required two rows of supports, one above the other, joined by vertical timbers (co-outacoués, étaux, gétos) and usually reinforced by wooden brackets underneath (poin plans) but later occasionally by iron rods or wire-rope suspension wires from above. The channel rested on the lower row. The whole structure was very skilfully adjusted and made fast with wedges and pegs and followed closely the irregularities of the rock face (figs. 8 and 9). The walk-way either made use of the protruding ends of the support beams (boutzets), sometimes with a plank on top of them and sometimes without (fig. 15), or consisted of a plank laid on the top of the channel itself (figs. 12 and 13).

Both systems necessitated the making of holes in the rock face. Often it was impossible to lower a man on a rope to do this because the rock face was too high or over-hanging. So a plank was pushed out into the void, often with a drop of hundreds of metres below, either from the last piece of solid ground on the route, or from the last support timber which had already been fixed in position, and its "shore-end" loaded with stones as a counter-balance. The workman then sat astride it, slid out along it to where the next hole had to be made, checked the

Fig 12: Niwärch. At "Z'leid Eggi", a series of Chänile (hollowed-out tree-trunks) with walkway on top, tied on with branches and creeper, the whole supported by Chrapfen.
(Historic photo).

Fig 13: Niwärch. A single Chänil with walk-way on top, supported by two Chrapfen.
(Photo: SAC Ortsgruppe Ausserberg)

Fig 14: Niwärch. At "Wiisu Flüe', walk-way on top of Chänil, the supports being apparently simple square beams wedged into holes made in the rock.   (Historic photo).

Fig 15: Bitscherin. Flowing 160m below and parallel to the Riederin, this suon carried water out of the Massa gorge to Bitsch-Mörel. The walk-way consisted solely of the protruding ends of the support beams of the channel.   (Photo: I. Mariétan)

gradient by one of the methods described above, laboriously made the hole with hammer and chisel and, finally and most hair-raising of all, fixed in it the next support timber – and so on. Sometimes, in the most dangerous locations, those who had been condemned to death were given this task. In a few places, both systems of construction were used side by side. An example of this existed in the Mund Gorge near Brig (fig. 16) with bisses of each type, one above the other.

Getting the original and replacement stores into position was often exceedingly difficult. For this reason areas of forest were reserved for use for the building and maintenance of nearby bisses, but this practice could introduce a danger: the cutting of trees tended to destabilise the

Fig 16: Mund. Superimposed water-courses in the Mundbach gorge. Drawing based on a 1920 sketch by L. Courthier.

shallow soil in places where this was most undesirable. Part-way along the suspended section of the Bisse de Savièse, a small saw-mill was set up so that timber requirements did not have to be carried further than absolutely necessary. At a point in the Baltschiedertal near Ausserberg, the only way to get the hollowed-out tree-trunks into position was to lower them from high above and the villagers had a heavy rope 200 metres long to do this.

The crossing of areas of scree, moraine, alluvial cones, bogs and other highly permeable ground presented the bisse builders with the problem of water loss by seepage through the bottom of the channel. In some cases wooden troughs were laid, either in or above the ground, across these stretches. Later most of these were replaced by galvanised iron or, still later, aluminium sections welded together (figs. 18 and 19) or cement, iron or, most recently, plastic pipes. Sometimes this loss of

Fig 17: Stigwasser. here shown cut out of a steep granite slope, carrying water from the Mundbach out of the Gredetschtal. *(Photo: I. Mariétan)*

water led to the abandonment of projects e.g. the Bisse des Audannes (Ohannes) built 1859-62 to draw water from the Lac des Audannes and abandoned in about 1870. In other cases, the builders had better luck because some solids in suspension help to seal the channels. A good example of this is the Bisse de Sion which originates at 1,800 where the waters from the Sixt des Eaux Froides join those from the Rocher d'Armeillon and which carry, early every spring, a particular cement-like alluvium which helps greatly to keep the bisse watertight. (If you go up above the Vauclusian springs below Armeillon to the strange Plan des

Fig 18: Grand Bisse d'Ayent. Metal channel carrying the bisse across unstable and porous ground near Le Samarin.

Fig 19: Bisse de Lentine. Metal channel carrying the bisse across porous ground near its prise d'eau on the Sionne.

Roses on the way to the Rawyl pass, you find a valley full of this alluvium which, I believe, soaks into the valley floor and comes out in the Vauclusian springs and so into the bisse). A very different man-made sealant used to be used in the annual repairs to the bisse de Savièse – see page 47 below. Wastage naturally varies greatly for reasons which, in the light of the foregoing, should be obvious; in some cases it accounted for 75% of the water put into a bisse at the prise d'eau.

In areas of sedimentary and stratiform rock or rock broken by naturally occurring splits, the rock face could sometimes be worked on so that the bisse could be entirely laid in the rock itself, e.g. Stigwassersuon in the Gredetschtal (fig. 17). Another good example of this is a section of the Bisse d'Ayent, where a soft stratum with the desired angle of declivity was found, from which it was possible to dig out a trench wide enough to allow the passage of a lot of water as well as to provide space for the gardien's walkway, all without using a single piece of wood. In some places a compromise was possible, e.g. in a sector of the Bisse de Savièse an L-shaped cut was made in the rock which provided the bed of the bisse and the wall on its "mountain" side, while the wall on its "valley" side was built of planks held in position by piles driven vertically into the rock (fig. 24). Where there was obvious danger of a bisse being damaged or blocked by avalanches, falling rocks or sliding soil, a protective cover (ploton) was made of trees laid longitudinally or logs laid transversely or, later, cement slabs – or the bisse was fed into a pipe.

The crossing of gorges and gulleys was effected by an extension of the methods already described. The wooden (both types) or later, metal sections of the channel were manoeuvred into position across the obstacle and the walk-way carried above the channel (figs. 20, 21 and 22).

The requirement for maintenance has always been continuous and costly. After the ravages of winter – snow, ice, rock falls, avalanches, and landslides – there is often a lot to be done in the spring before a bisse is ready to carry water – before the levée or mise en charge. The procedure which used to be an annual requirement in the case of the Bisse de Savièse before the Prabé tunnel was opened in 1935 was undoubtedly one of the most demanding because of the nature of that extraordinary beast; the procedure is described in detail in a very rare book of 1943 which may not be taken out of the Cantonal Library in Sion. Although this description is specific to Savièse, something like it must have been necessary to keep other bisses with similarly constructed sections in working order year after year. It is, therefore, of general interest and I summarise it as follows.

On a Sunday in April, the curé of St. Germain en Savièse announced in church that the work would begin on Monday week. The work entailed repairing all damage to the wooden channel and its supporting brackets, i.e. adjusting the planks, checking and, where necessary, replacing each piece of timber, calking all joints and cracks with moss or pine needles, clearing out and, where necessary, repairing the channel, both in the wooden runs (fig. 23) and in the areas where it was on solid ground. This usually took eight to ten days and was done under the orders of the "métral" and his five assistants, i.e. two procureurs, two arzieux and one chef des travaux. Every family was required to produce one or two workers and between one and two hundred men and women were available. After Mass, the métral organised the teams and detailed the tasks to be done. The women took shovels and cleared the ditches, stuffed the holes in low walls with moss and carried up arms full of white pine branches to caulk the joints between boutzets and

47

(Top) Fig 20: Oberstwasser. This was the highest and one of the oldest of the 9 suonen which carried water from the Mundbach out of the Gredetschtal. There was no walk-way, so the guard had to walk either in the water or with one foot on each rim of the channel. One is left wondering how it was ever built.
(Photo: I. Mariétan)

(Left) Fig 21: Grand Bisse d'Ayent. A relatively modern improvement whereby the bisse in its metal channel, with the walk-way on top, crosses the gorge of the Torrent de Forniri on a simple bridge made with two H-girders.

Fig 22: Bisse de Ro. The bisse crossed a gorge high above the Lienne by means of the "Passerelle des Amours". The walkway was minimal – and always wet when the bisse was flowing.   (Photo: A Vautrier)

planks. They also collected from the forest a type of light soil which they had learned to recognise wherever it appeared: it was a brown humus known as dajon composed of wood-dust, dry leaves and dead pine-needles and they placed big sacks of it at intervals along the wooden channel sections of the bisse.

When all these preliminaries had been completed, on a day chosen for the levée du bisse, everyone involved made their way to Ste. Marguerite's chapel, situated at the point where the bisse swings away from the Branlires precipice on to the Plateau of Savièse. Here Mass was celebrated and included a special prayer for those about to work on the

Fig 23: Bisse de Savièse. The repair party was carrying out the annual maintenance on the cantilevered channel across the Paroi des Branlires in 1934, for the last time; the 4,800m. Tunnel du Prabé was opened the following Spring. *(Photo Ch. Paris)*

bisse. Then everyone walked up it all the way to the prise d'eau on the Netage, where a large fire was lit and a substantial meal of raclette cheese and bread was eaten, irrigated with fendant, which produced high spirits and song, including the refrain which, translated from the patois, goes:

>Jamais Saviésan serait assez fou
>Pour s'en aller sans boire un coup
>(Never a Saviesan would be such an ass
>As to leave this spot without emptying a glass).

During this time the women filled the first 100 m. of the bisse with dajon. At a signal from the métral, the drainage sluice at the prise d'eau was shut and the water thus deflected into the bisse. One of the four vouasseurs (paddlers) then crouched in and blocked the channel with his body, to dam the head of the column of water, while the other three jumped into the ice-cold stream and stirred it to mix thoroughly the water and the dajon. This produced a sort of very fluid mud with which the entire channel was caulked. By sticking to the pine-needles with which all cracks and been filled, it stopped all leaks in a matter of minutes – and that was the secret of the operation. The bisse was not allowed to overflow and, as soon as the solution was evenly spread, the first vouasseur jumped out of the channel. Meanwhile, a second vouasseur had taken up position 50 metres downstream and the procedure was repeated at 50 metre intervals, with the vouasseurs always making sure to position themselves over a support bracket (poin plan) in case a plank at the bottom of the channel were to give way under the weight of the water. Where the bisse was wider, the efforts of three, or even all four of them were required to hold up the tide, which was called béra, the patois word for ram, used because the flow tended to be faster in the middle than at the sides, so that its front edge was reminiscent of a ram's horns.

The vouasseurs were, of course, soaked to the skin right from the start; after stuffing themselves with as much raclette as they could swallow, they spent three hours in ice-cold water. No-one complained; their forebears had done it for 500 years simply because it had to be done. The efficacy of the method was extraordinary. At first water trickled through every crack and space and crashed down the rock wall with a noise like a machine gun. After 30 seconds the noise subsided gradually and after a few minutes the section was totally watertight (fig. 24).

However, if anything failed, like one badly fixed plank, one loose boutzet, one piece of wall weakened by the winter, the result was disaster and, after repair, the entire operation had to be done again. But, assuming all to be well, as it was at that last levée in 1934, spirits rose higher and higher as the working party approached Ste. Marguerite's chapel. On 15th July, Ste. Marguerite's day, that year, a great feast was held outside the chapel and in Spring 1935 the Prabé tunnel, 4.8 km. long, was opened. The bisse delivers 700 or 800 litres per second, instead of its previous flow of 300. The water supply of the Savièse plateau, nearly all of which now comes through the tunnel to La Pétouse, is distributed mostly by pipe from there and is now plentiful. The system includes the old Dijore, Tsampi and Bourzi bisses which take water from the Drahin torrent from the Eastern end of the plateau; although still flowing (1993) they are no longer used for irrigation.

Some in Savièse who had business with their alpages near the Col de Sanetsch – and certainly some walkers up to or down from that col – must have regretted the abandonment and inevitable consequent disintegration of the old bisse because its walkway was a short-cut and the only way to avoid having to go down to Chandolin and up again.

The damaged suffered every year by some very exposed sections of some bisses led to the practice of building such sections so that they could be dismantled and stored safely throughout the winter. Maybe it was the analogy with this that gave the builders of the Furka-Oberalp railway the idea for their remarkable folding bridge over the Steffenbach gorge. The sudden arrival of storm water, at any time, could produce a flash flood, causing a bisse to overflow and burst its banks, not only cutting off the supply from its destination but also causing extensive damage below its course. This was a recurring element in the history of the bisse du Levron (see page 72).

The average life of the larch components of bisses was 25 to 30 years but some lasted much longer. The main timbers were usually marked with the initials of the family who installed them – or at least paid for their installation – and the dates when they were installed. On the recently-reconstructed section of the Grand Bisse d'Ayent above the Torrent Croix, some brackets dated 1814 were found still sound enough to use again. On the other hand, the notorious long wooden section of the Bisse de Savièse, referred to above, has totally disappeared. From the

Fig 24: Bisse de Savièse. the last "levée" of the bisse before the opening of the Tunnel du Prabé in 1935. Note how in this section the rock itself was used for the "mountain" side of the bisse and planks were used only for the "valley" side. (Photo: Ch Paris)

walkway along the Bisse de la Tsandra, on the opposite side of the Morge gorge, it was possible in 1993, with field glasses, to follow most of the former course of the Bisse de Savièse, but of all the great woodwork we could only spot one survivor - a plank which had somehow not been carried away.

Dimensions of bisses vary considerably, according to the amount of water available and to the desired capacity of the channel. A suon at Blatten (fig. 10) shows its hollowed-out tree trunks to be about 10 cms. wide, while 60 to 90 is a very rough average for width and 30-45 cms. for depth of the larger bisses like Ayent, Sion, Savièse, Varen, etc. In general, width is slightly greater than depth in order to facilitate the warming of the water. The flow of these larger bisses had been quoted as 300 to 400 litres a second but that leaves out of account the losses by seepage and evaporation referred to above. A 540 metre tunnel on Bitscherin – a replacement for a suspended wooden section which had caused a number of deaths – was made wide enough to serve also as a flume for timber from the forest higher up the mountain and the 1914 regulations of the Grand Bisse de la Rioutaz at Lens also provide for the floating down it of timber, which could not be longer than 50 cms unless special permission has been obtained beforehand. There are stories, probably apocryphal, of goods being moved in barrels on a section replaced long ago of a suon at Visperterminen.

In some places bisses are superimposed, one above the other, on the mountain side. The Val de Nendaz, the Mayens de Sion, the West side of the Val d'Hérémence, the Mayens de Réchy, the Gredetschtal (Mundbach) (fig. 16) and the Baltschiedertal have good examples of this (see maps on pages 112 & 115). Writing in 1911, Dr. Louis Lehmann suggested that the advantages of such arrangements included the avoidance of disagreements between several communes all wanting water from the same source at the same time and the reduction of the danger of the ground slipping because relatively unstable terrain supported several small channels better than one big one: in addition, the building of small channels was simpler and cheaper and watering was more convenient and practicable, specially in the spring because upper and lower areas do not have to be watered at the same time. These factors may have led to situations like that in the middle of the Gredetschtal (Mundbach) where, on the right bank, six bisses were built within about 25 metres of each other and on the left, three, or the

Baltschiedertal with four on each side. Such arrangements seem nowadays unnecessarily hazardous and costly to maintain but, of course, no alternatives – such as a tunnel or one big pipe with a suitable switching system – were available to provide essential water at the time these bisses were built and so, inevitably, each local requirement was treated separately. However, those arguments alone cannot account for the existence of four suonen running parallel, only about 150 metres apart in height, through relatively easy terrain of woodland and cultivation to bring water from the glacier-fed Riedbach to the dry plateau of Grächen above the Mattervisp valley. Ignace Mariétan, writing in 1948, rather dismissively put this apparently unnecessary complication down to the eccentricity of the local inhabitants. A similar situation existed in Oberried until 1937 when three parallel suonen, only 50 and 150 metres apart, were replaced by a tunnel.

In the courses of some of the longer bisses, it was found necessary to reinforce by mechanical means the supervision provided by the gardien's daily perambulation. This took the form of the barrate, (tourniquet avertisseur, moulinet, Wasserschlegel, Merkhammer or Wasserhammer) consisting essentially of a paddle-wheel mounted in the centre of a restricted section of the bisse and turned by the flowing water (fig. 25). By a simple crank on one end of its axle, it lifted a wooden hammer which, on falling down again, struck a wooden sounding-board. The noise of these regular blows could be heard for considerable distances; if it stopped, the gardien knew that the flow of water was interrupted and set out with all speed to locate and repair the trouble. We found a small, modern, all-metal version of the Wasserhammer on the Niwärch above Ausserberg in 1994; it was not fool-proof, however, because a broken branch had got stuck in it and jammed it into silence, but the water was still flowing (fig. **6**).

At the point where the water brought by the bisse has to be divided among those entitled to it, a distributor (répartiteur or diviseur) is installed. In its original but very effective form, this consists of a stout wooden box, about a 1 metre cube, with as many square or rectangular holes in its side walls as there are channels to be served (figs. 26 and 27). These holes are all the same size and at the same level in the sides of the box. The word "bulletin" is used variously to mean the hole, the flow of water in the channel, or a water-right. The water from the bisse is fed into the box at a higher level than the bottom of the holes. The area

Fig 25: Water-hammer: the device for making an audible signal – the blows of the crank-operated hammer on its sounding-board – to indicate that the flow of the bisse had not been interrupted.

which each channel has to irrigate determines its quota of so many bulletins (in this case meaning holes) per hour and although the amount received depends on the pressure of the flow and is, therefore, not a fixed unit, it is a fixed fraction of the total and thus ensures that the supply is distributed in an equitable manner. The box also serves as a sand-trap which can easily be emptied from time to time. An example of a later, more sophisticated combined prise d'eau and répartiteur was built in 1986 at les Yvouettes (Dorbon) on the Sionne (fig. 7). This is a substantial installation built of concrete with iron grilles; it not only divides the water between Sion ($1/4$), Ayent and Arbaz (together $3/4$, subsequently divided again $2/3$ to Ayent and $1/3$ to Arbaz) (fig. 28), but also prevents branches and other debris getting into the bisses concerned.

Distribution from main to subsidiary channels is usually by wooden or metal sluices held in small metal or concrete mountings. Many metal ones have padlocks on their handles so that they cannot be interfered with. The final distribution by each entitled party of his share of water on to his own land is normally left to him and is carried out with two simple tools, known variously as tranchant, étanche, torniou, or Wässerplatte, and pioche, délabre, Wässerbeil, or Wässerhaue, respectively (fig. 29). The iron étanche or Wässerplatte is pushed into the bed of the bisse, diagonally to the flow, and the pioche or Wässerhaue is used to cut a channel through the bank, just upstream of the étanche to divert the water through the cut into a field or minor channel. Sometimes a flat stone is used instead of the étanche. The pioche is used again to restore the bank securely when the quota of water has flowed.

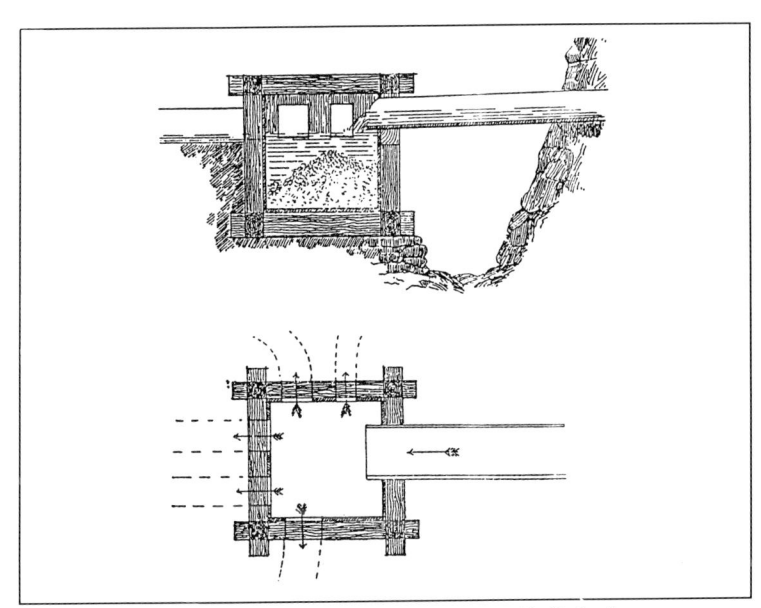

Fig 26: Sand-trap and distributor with 5 "bulletins".

Fig 27: 4-way distributor.

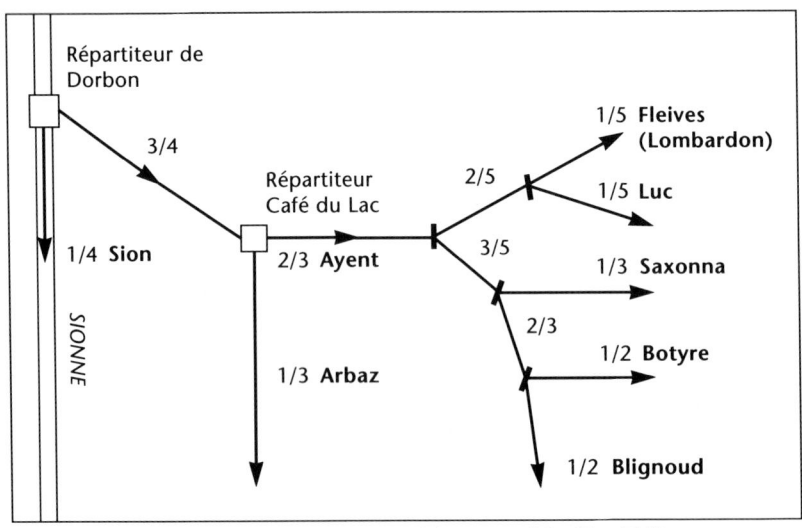

Fig 28: Diagram of the division of the water of the Bisse Taillaz.

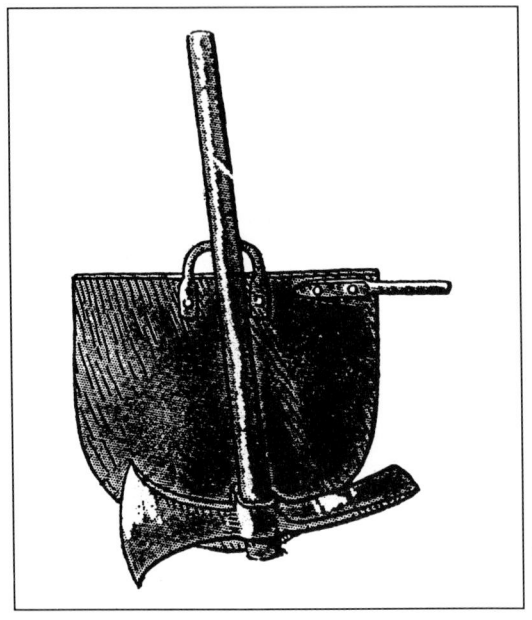

Fig 29: Étanche (Wässerplatte) and Pioche (Wässerbeil)

Chapter 4

Administration and Regulations

Just as the problems posed by the construction of each bisse were, inevitably, sui generis, so the regulations for administration and operation of each one have been developed ad hoc. All these regulations (of which there are obviously numerous variants) share two characteristics: first, very precise rights and duties and, secondly, the enforcement of duties by threats of dire punishment. The regulations of the Bitailla in 1306 called for the cutting off of a hand of anyone who stole someone else's water. Later, substantial fines were imposed – ibid. 1621, 1763, 1771 and so on, and smaller fines for minor offences, such as taking water for too long. The strictness of enforcement was in practice – and still is – directly related to the hazards of supply.

In the earliest days, the regulations of all bisses were by custom but, given the opportunities for argument and strife of all sorts between hard-pressed communities, it is not surprising that codification, in Latin, came quite soon, e.g. for Bitailla (Ayent) in 1306; Ausserberg 1346; Alte (Turtmanntal) in 1381; Rioutaz in 1394 – specifically stating that the regulations put on record what had already been the practice for 33 years – ; Grand Bisse d'Ayent, 1448; Bisse du Levron, 1465. Translations into the vernacular, at first very rare, began towards the end of the 16th century. Every year until 1967 the commune of Grimisuat had to ask Sion for permission to take its share of the water of the Sionne, quoting an Act of 1715. In June 1945 Grimisuat instructed its guard to make his rounds on Mondays, Wednesdays and Saturdays and expressed the hope that Sion would do the same for Tuesdays, Thursdays and Saturdays (why Saturday should have a double inspection is not clear).

There are three types of body controlling and administering bisses: first, associations (consortages/Suonengenossenschaften) of those concerned in one commune when the bisse was built exclusively for their land; secondly, associations of those concerned in several communes when the bisse was built for their joint use, e.g Ausserberg and Gründen; thirdly – and until recently more rarely – municipalities, when the bisse was built at the expense of the commune, e.g. Sion and Saxon. As in so many aspects of this subject, for historical reasons there is no standardisation and complications abound. Ayent and Grimisuat are joint owners but Ayent negotiates as a consortage and Grimisuat as a commune. The Bisse de Savièse was originally a consortage and the consortage still exists to distribute the water, but the Prabé tunnel, opened in 1935, belongs to the commune and the commune negotiates with outsiders – and exceptionally restrictively. Rights cannot be sold outside the commune and, except for the water sold and piped to Grimisuat, water cannot be used to irrigate land which members of the commune own outside it. In the case of Bitailla, the water-rights are attached to the land (meadows have no claim to water but arable fields do) and are inherited, bought and sold with it; in the case of the Grand Bisse d'Ayent, they are inherited, bought and sold separately, but they cannot be acquired by anyone who is not a "bourgeois" of Ayent.

Some bisses are still governed by regulations made 400 or more years ago; in other cases regulations have been reviewed and brought up-to-date. For example, detailed regulations for the Grande Bisse de Lens or Bisse de la Rioutaz, based on its division into twelve areas of responsibility, were drawn up on 18 October 1457. The first revision was made in 1698, the second in 1914 and the third in 1980. The following details are an abbreviation of the 1914 revision, which was comprehensive and at the same time fairly typical in effect, even if not obviously so in form, of those of many other bisses.

> a) The Grand Bisse is divided into four Quarters: Lens, Chermignon, Icogne and Montana with Chermignon-d'en-Bas; 700 individuals are entitled to water rights.
>
> b) Each quarter is divided into sections called "Tassoz", the upkeep of which is the sole responsibility of the owners of the water of that quarter.
>
> c) The cost of the upkeep of the prise d'eau (the source of the bisse) at the place called Couchellet under the Barmes rock on the Lienne river is the responsibility of the communes of Lens, Chermignon, Montana and Icogne.

d) The consorts of the Bisse undertake to keep it flowing until the end of November for the watering of cattle, the supply of water-troughs and for fire-fighting. The boundaries of 15 Tassoz, with their lengths (ranging from 17 to 3434 metres) and of the four quarters (totalling 13,744 metres) are set out.

e) The responsibility for the sluices, some belonging to the Bisse itself and the rest to individual authorised parties, is detailed.

f) The administration of the Bisse is composed of three different bodies, each with distinct functions:

> i) the Superior Authority is the Commission Grande Bourgeoisiale and, failing it, the Councils of the four communes (Lens, Chermignon, Icogne and Montana with Chermignon-d'en Bas). It supervises the works and general interest of the Bisse.
>
> ii) the Councils of the four Communes each undertake the works ordered by the Commission du Bisse: each receives the annual accounts of the mignour (see below) of the expenditure made in each accounting period; fixes the daily rate of pay and the amounts to be paid by each entitled party; initiates court action, in the name of the Quarter, against anyone who has damaged the Bisse or its interest within the limits of its Tassoz.
>
> iii) the Commission du Bisse is composed of the avoyour (president) and the four mignours.

g) The avoyour is appointed for four years and is selected from the four communes in rotation. His functions are:

> i) to supervise the mignours and the guard;
>
> ii) to oversee and generally attend to the interests of the Bisse;
>
> iii) to present to the Commission du Bisse the candidate for the appointment as guard;
>
> iv) to arrange for the execution of urgent works by third parties, at the expense of any Quarter which has failed to do so itself;
>
> v) to report to the Superior Authority in the event of dispute;
>
> vi) to fix the date of the annual inspection of the Bisse.

h) The Council of each commune appoints for four years the mignour of its respective Quarter. His functions are:

> i) to select the men required for annual and special work on the Bisse;
>
> ii) to supervise all such work and keep accounts for it and keep an eye open for anything that needs doing urgently;
>
> iii) to submit an annual account to the Administration of his commune or Quarter;
>
> iv) to sell the Meschuires (see below) and collect the proceeds;
>
> v) to supervise the flow of water from

the first sluice at which it enters his area and to ensure its continued passage out of his area at the proper time.

i) The guard is appointed each year by the Commission. He has to be sound, zealous and reliable and has to provide a financial deposit on appointment. Applicants for the post apply to the avoyour quoting the salary required. The post cannot be combined with that of avoyour or mignour and the man appointed cannot delegate his functions to a substitute except in an emergency, which must be referred to the Commission du Bisse. The guard's functions are:

> i) to watch over the operation of the Bisse – to ensure that the maximum possible quantity of water flows during its periods of operation;

> ii) to walk the length of and inspect half the Bisse every day and to make as many other inspections as may be necessary in times of frost, rain, hail or showers, that is to say whenever there is danger of a break in the Bisse or a serious loss of water for the consorts;

> iii) to carry out, free of charge, not more than two hours work in any Quarter. Additional work is paid at an hourly rate and he must warn the mignours about requirements for urgent repairs;

> iv) to report to each mignour any damage or breakage of the rules perpetuated in his respective Quarter;

> v) to pay for a Low Mass on the day of levée (putting into service) of the bisse, for the relief of souls in Purgatory.

j) The four Quarters are managed by the communal administrations. Each Quarter is entitled to 13 "poses" (watering periods) of which the total duration is $3^1/_3$ days, so there are 3 poses per 24 hours. The morning one begins at sunrise and finishes at midday; the afternoon one runs from midday until sunset; the night one from sunset until sunrise.

k) "Componction" is the right to extend one watering period into the next if the bisse is not flowing at full capacity. The avoyour, the four mignours and the guard have certain duties to watch for and report to the Superior Authority any misdeed which is interfering with the flow of the bisse. Three poses de St. Jacques and the Meschuires are short watering periods which can be sold by the mignour at a price which must not be below that paid by those entitled to water-rights. The proceeds are divided equally between the Church, the mignour himself and a recipient chosen by the relevant Quarter.

l) The Bisse must be repaired every year in time for its flow to start before the feast of St. George the Martyr, unless there is too much snow or a storm or everyone agrees that repairs should be postponed because there is too much water. If any Quarter does not put work in hand in its area in time, the avoyour must put it out to contract at the expense of the Quarter concerned.

To complete the history of the regulations of this bisse, a few words are needed about the third revision, in 1980. This was felt necessary because of changes in the pattern of the agricultural and viticultural use of the land. The number of fields grew fewer, as did the number of cattle; the construction of water storage basins and a network of pipes in the different communes enabled the vineyards to be irrigated by aspersion, i.e. overhead sprinklers, instead of by ruissellage, i.e. by causing water to run on the surface of the ground, which inevitably damages it by erosion. This change pointed up the need to modify the administration of the Grande Bisse. After 530 years of administration "grand bourgeoisiale" , the Grande Bisse was handed over to the consorts, to be administered by them, with the aid of a committee of 13 members, 9 of whom represent the consorts and the communes.

The organisation of the Bisse de Savièse is rather different from that of Lens. It serves six zones, inappropriately called Quarters: Moussy, Syllandau, Huchelet, Montana, Rocher and Arvisy. The unit of water is the "pose", which consists in this case of the flow of water in a Quarter for three hours. There are in total 840 poses. Each Quarter has a pond (gouille), to accumulate the water flowing on Sundays and holidays because on those days no watering is done (unlike in some other areas, where dispensation has been obtained from the bishop to allow Sunday afternoons to be worked). The supervisory authority consists of the computiste (the "figures man") and his assistant, the métral (foreman), the procureurs (assistant supervisors) and the azieux (waterers).

In general, where the administering authority is a consortage it usually meets twice a year, in Spring and Autumn, and constitutes itself a general assembly. It is to these meetings that all demands concerning the acquisition of water or its transfer to another piece of land have to be submitted. Some consortages can sell water to farmers who are not members of the consortage, but only if a majority of the consorts agree. Each consort has a vote at these assemblies for the election of the president, the procureurs (Vögte in German-speaking areas) and the guards (Hüter). The president or directeur of the bisse is the organiser of all work connected with it: he fixes the dates of repairs, the starting date of watering and, if anyone has stolen water, the president imposes the appropriate fine and informs the judge about the offender. The procureurs record in the Livret d'Eau (Wasserbüchlein) the droits d'eau (water-rights) (confusingly, in some areas called bulletins) allocated to each user; each droit allows so many hours watering during the summer and anyone who is entitled to two or more droits is required to be a procureur for a year. For the secondary bisses – branch distribution channels – there is simply a commandeur d'eau. At Lax (Goms), where d'Bärgerwissa (Laxerwyssa) is controlled by the municipality, water discussions take place under the chairmanship of the mayor and the procureurs carry out all the executive functions, including the appointment of the guards. Some of the longer bisses had several guards (Saxon had four), each of whom lived in a little hut beside or over the bisse for the duration of the watering season, usually about four months. Located near each hut was a water-hammer, referred to above, to give warning if the flow was interrupted. Some committees apparently did not entirely trust their guards to do their job conscientiously. At Niederwald (Goms) each guard had to hang up a tally at the end of his walk of inspection and the next day his colleague had to bring it back to the village. On the other hand, on some bisses flowing through more serene country, where the danger of breakages was considered small, simple flow-measuring devices were installed. If, when the guard on his tour of inspection reached the device, he could see that the water-level had not dropped, he could assume that no loss was occurring above that point and the rest of his walk could be dispensed with.

The water-rights possessed by any individual interested party are often hundreds of years old; they may originate in the participation of the original owners in the building of the bisse. In the course of centuries, properties have been broken up, sometimes by inheritance, sometimes by sale; further complications have been occasioned by an owner selling land without selling the water-rights, a practice possible in some communes but not in others. The building of the Bern-Lötschberg-Simplon railway through the lands of Eggerberg and Ausserberg necessitated a revision of the water-rights; on the other hand in some areas, such revisions have always happened at intervals of 10 or 20 years to prevent certain privileged properties permanently taking water to the detriment of less fortunate ones. At Visperterminen a commission reviewed and renewed water-rights every year on 21 September. At Saxon great exactitude of distribution was achieved by the directeur of each channel fed by the bisse distributing, 24 hours in advance, a bon d'irrigation (watering ticket) for every 2 hours flow per 500 square metres. In general, the shorter the water supply, the more exact the distribution and the tighter the regulations: the difference is apparent even in neighbouring communes, such as Fiesch and Niederwald. As a typical example of watering regulations, here is a summary of Dr. Louis Lehmann's (1911) version of the rules of the commune of Saint-Luc:

> Art. 1 – Every owner of meadows in the commune of Saint-Luc has a right to irrigation water.
>
> Art. 2 – The upkeep of the lower, new and upper bisses, the mill-leat to the Zariré canal, the Zariré canal and the bisse from Zariré to la Barmaz are all the responsibility of the municipality,
>
> Art. 3 – Other secondary bisses are the responsibility of

their owners.

Art. 4 – The distribution of the waters of the three main bisses is reviewed every ten years; the same applies to the secondary bisses if those entitled so wish.

Art. 5 – Each year the communal administration will announce at the ordinary public meetings the date agreed for the start of watering. Before that date watering is free. However, he who has the water first keeps it for the time necessary to water his fields in accordance with current rules.

Art. 6 – Whoever uses the water of the three main bisses is required to divert it only at the approved places. Failure to do so incurs a fine of 2 francs. Anyone who carries out this task badly will be fined 1 franc; any damage resulting from the offender's action will be at his expense; the same applies to the secondary bisses.

Art. 7 – As a precautionary measure against fire, the last to irrigate with the water of the main bisse is required, on finishing, to redirect the flow of the said bisse to Zariré. Failure to do so incurs a fine of 50 centimes. Furthermore, he will have to pay the cost of the procureurs' tea if their intervention is necessary to sort things out.

Art. 8 – All those who irrigate with water from the main bisse are forbidden to let the water run to waste in the déchargeoir (drainage channel) of Amourné except for such time as it takes to move from one property to any other or to come to Zariré; in all other cases it must be caused to flow to the last-named. Contravention of this article will incur a fine of 2 francs and the cost of any resultant damage.

Art. 9 – All who "abandonne l'eau en couchée" (I have not been able to find out what that means) on their property will incur a fine of 2 francs by day and 4 by night and the cost of any resultant damage.

Art. 10 – Anyone who steals water from another will incur a fine of 1 franc plus damages according to the seriousness of the offence.

Art. 11 – All those entitled to water on the same day are required to meet at 8 o'clock on the previous evening to agree the order of their turns. Anyone who fails to appear will be put last.

Art. 12 – If the flow is not started in the morning or if there is an interruption during the day, anyone who has not complied with Art. 11 cannot claim any water until all those who arrived as required have finished watering, irrespective of his allocated time.

Art. 13 – Watering at night is forbidden at Amourné or between the reservoir there and the déchargeoir of the Eghvogerts. Contravention incurs a fine of 4 francs.

Art. 14 – Watering in the commune is forbidden on Sundays and holidays. Contravention incurs a fine of 5 francs.

The style of many of the bisse regulations is robust and simple but by no means always grammatical or clear to the uninitiated. The regulations for the bisse de Saxon on the subjects of Saint-Luc's Art. 6 and 7 are better:

Art. 6 – The owner of a bulletin (meaning in this case a droit d'eau) can use it as he wishes, provided this does no harm to his neighbour whose turn is next. After use, he must restore the water to its normal course as soon as possible. Omission to do so incurs a fine of 15 francs.

Art. 7 – As soon as his appointed watering time has elapsed, a proprietor must close and block all the openings which he has made in the side of the bisse to let the water run into his property, so that the bisse receives the total flow of water and lets it pass onwards without interruption. Omission to do so incurs a fine.

Chapter 5

Disputes

With the many expressions in the records of good neighbourliness ("à titre de bon voisinage et de solidarité"), it might appear that all was sweetness and light in intercommunal relations about water, but deeper examination reveals that this was often far from being the case. In some areas, claims and counter-claims were endemic. If disputes could not be settled locally, they were referred to the Bishop of Sion who was, as mentioned in Chapter 2, the temporal as well as the ecclesiastical authority of the canton until the Rénonciation à la Caroline in 1627. Thereafter disputes were handled by the cantonal civil courts, but, as local priests have always played a most important role in the everyday lives of Valaisans, it was inevitable that they remained involved with bisses. Water as an essential to life was obviously holy and it and everything to do with it had frequently to be blessed; the Almighty had to be asked to intercede to ensure its supply, to preserve the lives of those who worked on bisses and, if He sometimes failed to do that, to provide salvation for the souls of the casualties and solace for their families and friends. Priests went out with their parishioners to work on bisses (fig. 8) and to resolve any problems which arose in the course of labours which, in the nature of things, were near to God.

At the risk of invidiously omitting many other devoted clerics, I mention one medieval one and five modern ones: Prior Jean de Lens, who built the Bisse de la Rioutaz in only two years, 1448-1450; the Rev. Curé Cordel of Ayent who rallied his flock to repair the Grand Bisse d'Ayent when, at the beginning of the 19th century, it was battered into uselessness by a series of avalanches, rock-falls and all manner of bad luck and he wrote songs for them to sing to encourage them in their labours; the Rev. Canon Fardel, a conscientious historian of everyone's achievements except his own, rebuilt the parish church of St. Romain in 1854 after personally applying to the Pope for permission to draw on central church funds to do so; he was a tough, authoritarian, independent-minded priest who was nicknamed "The Pope of Ayent";

the Rev. Curé Pierre Jean of Savièse, who officiated during the last years of the old Bisse de Savièse and duly blessed the tunnel which replaced it; the Rev Canon Lucien Quaglia of Lens, who produced a masterly history – water disputes and all – of Lens; the Rev. P. Sulpice Crettaz de la Villaz, who did the same for Ayent.

In general, it seems – naturally enough – that there was more trouble where bisses served or in some way involved two or more communes than where consortages were concerned. Lax and Martisberg were both supplied with water by the Deischbach and by the Bärgerwissa suon, once 17.5 km. long, and drawing water from the Western snout of the Fieschergletscher, but, after that glacier receded, appreciably shorter and taking water from the Wannigletscher. Both claimed the water as their own. This quarrel was recorded in 1347 and again in 1367. It was eventually agreed that Martisberg should take the water from Easter to the feast of St. Peter and St. Paul ("without argument or hindrance") and that Lax should have it from then until the following "Maria Geburt"; from then until Easter it should be shared. The matter was taken twice to the Bishop of Sion to give a ruling on some aspect of this quarrel in 1443 and 1554. Thereafter, apparently, there was peace but in the meantime trouble had developed between Fiesch and Fieschertal of the one part and Martisberg of the other about water from the Mattenbach, Mittelbach and Wysbach. In 1351 an agreement had been reached at the church of St. John the Baptist at Fiesch to build and maintain a new suon; further details were finally agreed in 1747. Argument arose again in Napoleonic times, when the Emperor regarded Valais as part of his strategic route to Italy; a document recording the decisions reached in 1811 is headed "Französisches Reich, Department des Simpelberges" and was still operative in 1961.

Trouble blew up between Ayent and Lens in 1448 when the Lensards were starting to build their Grand Bisse (or Bisse de la Rioutaz). The Ayentots objected to the location chosen for its prise d'eau from the Lienne. It was eventually decided to settle the matter by a duel between the champions of each commune. Ayent's man was an invincible-looking giant; Lens's man was a crafty little fellow who watched his opponent very carefully, caught him off his guard and threw him to the ground with a long length of wild vine (rioutaz); hence the alternative name of this bisse, used to this day. Argument between these two communes, on opposite sides of the Lienne seemed to have rumbled on while, in contrast, Salgesch and Varen on one bank of the Raspille and Miège and Sierre, served by the Bisse de Zittoret, on the other managed to get on without any serious strife. Either the fact that the Raspille is the language frontier, thus making argument more difficult, or a more understanding and less parochial attitude on the part of those concerned may have helped. Vex and Hérémence were in dispute for half a century over the shares of and rights over the bisse known as the Grand Trait. Vex claimed that it did not receive enough water and then Hérémence maintained that any enlargement of the bisse would constitute a flood danger to its lands. In 1484 there was a three-cornered quarrel when the Sédunois (people of Sion) accused the Ayentots of stealing the water of the Sionne; the Ayentots said that all they had done was to take into their bisse the water from La Combaz which had always belonged to them anyway and at least since 1400 and, furthermore, that the Sédunois should be grateful to them for reducing the flow of the Sionne when it threatened to overflow. The Sédunois then accused the Ayentots of selling water to Savièse; the Ayentots denied it, saying that, anyway, quarrels between Sion and Savièse were no concern of theirs, so the Sédunois then accused Savièse – and so on. La Combaz figured again in 1911 and 1912 when the people of Arbaz built themselves a bisse to carry off its water; this provoked such uproar from Ayent that the project was dropped and the bisse remained "sec". However, in the mid 1980s two new underground springs were found, one on each side of the Sionne above the Répartiteur de Dorbon. These were not used, as I first assumed, to increase the flow of the Sionne but piped away as a new drinking water supply for Arbaz, so perhaps the Arbaziens had the last word in this dispute. Another disagreement arose in 1686 when the people of Arbaz put water from the Bitailla into the Grand Bisse d'Ayent and took water out further down. It is not clear whether the Arbaziens were being dishonest but the people of Grimisuat reckoned that they were being wronged; they protested and demanded that their rights be respected; the Bishop decided that the Arbaziens could use the Grand Bisse provided that they built a partition in it to keep their water separate from that of Grimisuat, the rights to which had been bought by the latter in 1464. No trace of that partition remains, but there is an arrangement to switch some Bitailla water into the Bisse d'Ayent in the event of an emergency. Several hundred years later a similar arrangement was made concerning the Bisse de Saxon, into which water from the Faraz torrent was put to be taken out lower down and put into the Ecône torrent to increase its flow and provide irrigation for some of the land of the commune of Riddes. Another area of frequent if not continuous disputes involved the Bisses of Grône, Loye, Vercorin, Chalais and Réchy which take their water from the Navisence and Réchy torrents. A succession of disputes in this area, especially in the 16th and early 17th centuries, was referred to the ecclesiastical authorities.

However, the fiercest dispute, or at least the best-documented one, seems to have been that between Le Levron and other villages in the Valley of Bagnes. The people of Le Levron asked the Abbot of Saint-Maurice in 1443 for authority to take water from Chardonnay to irrigate their lands. It seems that nothing happened until the Levroniens applied again in 1465. The people of Sarreyer, Montagnier and Cotterg opposed the project from the start. By 1471 the consortage was legally established but in 1478 this bisse was destroyed by the people of Bagnes. After reconstruction, the dispute continued on and off, notably in 1515, 1545, 1626, 1629/30 and 1839, until the Bisse du Levron was abandoned in 1923. In this case the quarrel was not only about rights to water; the opposition claimed, as in the case of Hérémence mentioned above, that the bisse would damage their lands by causing flooding and landslides when it overflowed or was broken by storms or rock-falls. That this claim was not without substance is proved by the amounts of the costs which the Levroniens had to pay for just such accidents over the years – and this in spite of the appointment of "gardes sur Bagnes", to ensure that no damage to that commune was caused, and the presence after 1839 of representatives of Bagnes at each formal closing of the bisse. With the wisdom of hindsight, Clément Bérard, writing in the sixties, includes a quote from someone, whom he does not name, to the effect that the water of Chardonnày, had it been wisely collected and distributed, would have sufficed in normal times to fill the three Bisses of Le Levron, Sarreyer and Verbier/Montagnier without the disputes which so often arose. There must have been many other cases elsewhere in the Canton where parochial obstinacy led to discord, duplication, waste and a regrettable tarnishing of the remarkable achievements of many generations of admirable Valaisans. Strains and stresses involving the Bisse de la Tsandra and the resolution of them by the Conseil d'Etat are detailed on page 95.

As has been explained in Chapter 4, the regulations governing bisses aimed at the prevention of misuse of water. Provisions were made to dissuade potential offenders from taking water to which they were not entitled and from taking too much by taking it for too long. As the Middle Ages progressed the population grew, in spite of losses caused by recurring outbreaks of plague, and the profitability of the export of cattle to Italy led to an increase in the number of cattle and a decrease in the number of sheep and goats; this was, in effect, a commercial development superimposed on the natural scene – and it needed more water so that the maximum possible number of thirstier animals could be reared. This tended to produce disputes between individuals rather than between communes or consortages and, in a deeply religious community, condigne punishments, such as long terms in Purgatory, were threatened and feared and figure in the folk-lore as, for example, in the recurring theme of ghostly processions, especially at night near glaciers, of those who in their lives stole water and did not atone for it by repairing the damage which they had done.

Fig 30: Bâton à Marques. One method of recording the ownership of rights to the distribution of water. This example was used for the Bisse de Savièse from 1841 to 1860.   (Photo: I. Mariétan)

CHAPTER 6

Water Rights and Irrigation

Just as detailed regulations were necessarily made for the bisse organisations themselves, so similar care had to be given to the control of the water-rights and actual watering procedures. (Surprisingly, no written regulations appear to have been made for the use of the water of the Bisse du Levron, those concerned having apparently managed to get on perfectly well without them.)

For much of the long life-time of the bisses, few people could read or write, so that a written index of who owned what in the way of water-rights was out of the question. In some areas each family possessing water-rights had its badge or sign, which consisted of an individual arrangement of lines and dots. The badges were carried on one side of 1 to 1^1/$_2$ metre long pieces of wood and on the opposite side the number of water-rights was carved in Roman numerals. This device was called a bâton à marques, Wasserscheit or Wasserknebel; the last one for the Bisse de Savièse was made in 1841 and remained in use until 1860. It contained 269 badges and 1092 water rights (fig. 30).

Another method of recording rights was the Wassertessle or Tessel (tally), a wooden block on which the number of hours watering was marked. These were kept by the procureurs and still used in the 1920s in Mund, Zeneggen and in the Lötschental and until 1933 in Chermignon. Each entitled proprietor had one, on which his house badge and quantity of rights were shown. The tessel was notched in the middle, one end indicating morning and the other afternoon. Every morning during the watering period (see below) the erwin (official in charge of water allocation) hung one up outside the house of each of those entitled to take water on that day, with morning or afternoon sign uppermost, as appropriate. The division of time for the purpose of calculating water-rights took various forms. In some areas the day was divided into quarters (04.00 to 09.00 (morning); 09.00 to 14.00 (mid-day); 14.00 to 20.00 (vespers); 20.00 to 04.00 (night). In others, the hour was divided into periods ranging down to $1/8$ hour. As late as 1950, in some areas the fixing of times was not done by the clock but by the position of the sun and shadows, e.g. in Törbel "Wissbergschiene" meant the time when the first rays of sunlight hit the Weisshorn, while in Staldenried "Dreifurrenschiene" meant the moment when the same happened to the Dreifurren, the highest field above the village. Similarly, in the evening "Schattengspon" was the moment when the shadow of the mountain reached the first meadows of Gspon.

Irrigation is an art as well as a science. One needs to know the quantity of water required at each watering and, indeed, the ideal manner of the watering. The ground is conditioned to the amount of water regularly received each year; if the amount is too small, irregular or too great, the effect is damaging, though, by and large, a little too much and too often is better than the opposite. Watering must never be left to itself but always supervised to avoid accidents and many of the locals insist that one should never be alone when irrigating from a bisse. One season's carelessness can ruin several years' harvests. The worries about damage and costs of floods caused by bisses over-flowing or breaking their banks are constantly recurring themes in their history. The importance of the silt carried in suspension in bisse water has already been mentioned; bisse water is also diverted to carry manure from stables and enclosures out on to the land. Manure thus carried in diluted form through small channels penetrates better into the soil than that transported by other methods.

The presence of a bisse is obvious, even from a considerable distance, because of the difference of colour and lushness of the vegetation, both in the fields which the bisse irrigates and along its course (fig. 7). Even very young animals react to this almost as soon as they can move (fig. 31).

Watering is done when the land requires it and the needs of meadows and vines are different. The former, often watered day and night, are irrigated four or five times a year, two or three times for the first grass crop and once or twice for the second. Watering by night has the advantage that the loss by evaporation is less. Land in the shade needs less water than that in the sun and gravelly soil more than clay – at Raron, for example, land has to be watered once a fortnight.

Vines, on the other hand, are usually watered twice during the summer; first, when they begin to flower, about mid-June, and, secondly, when

Fig 31: Niwärch. The presence of the bisse is revealed by lusher wild vegetation as well as irrigated land. The youngest local inhabitant, who was only just able to stand up, had already found the best spot on which to lie.

the grapes begin to ripen in August. If the ground becomes exceptionally dry, a third watering takes place. Formerly – and indeed in some areas still – watering vines involves directing the water into little channels along the wall at the top of the "tablats" (terraces) to a series of small holes in the banks of the channels through which it flows down among the vines ("ruissellage" or "ruissellement"). Where the ground is very stony, thus impeding the flow, or where the tablat is very large, movable metal channels are used to direct the water to three rows of vines at a time. One starts at the bottom and gradually works upwards by removing the sections of channel one by one. The whole operation must be very carefully controlled because, if the ground is allowed to become too wet, the walls disintegrate, the terraces collapse and the soil is washed away. Nowadays, when irrigation can be carried out by overhead sprinkler (aspersion), this danger can be avoided, but it requires a sufficient difference in levels to produce the head of water necessary to operate the system.

These differing requirements have to be coordinated with the operating periods of each bisse. Although for most bisses there is an agreed fixed starting date, it does not follow that the watering will begin on that date, but it fixes the "tour" (Wasserkehr), the rotation in which those entitled take their water. The length of the tour is the time it takes to water all the land which the bisse serves, for example, at Ausserberg, tours of Niwärch were 21 days, of Mittla 18, of Undra 21 and of Manera 16. Starting dates vary considerably. A substantial number, at both ends of the Canton, are recorded as beginning in February or March, which does not accord with Dr. Louis Lehmann's assertion that in the Haut Valais it is rare to start before late April or early May. Watering usually entails a number of periods, each of a fixed number of days, for example, the Bisse de Riccard, now largely in a tunnel, draws water from the Navisence and delivers to Chalais 400 litres per second from April in 7 or 8 periods each of 18 days; the nine bisses from the Mundbach serving land between Lalden and Mund and between Birgisch and Naters operate for 8 to 10 periods each of 8 to 12 days; the two from the Massa serving Bietsch and Ried-Mörel, now in tunnels, operate from late April for 10 periods each of 14 days. These periods may, in case of need, be varied with the weather.

Watering stops in mid-August except in the hottest summers, when it may be extended until the end of the month. At St-Luc and Vissoye, on the sunny side of the Val d'Anniviers, watering starts a fortnight before doing so at Grimenz, on the shady side. Lower areas are watered earlier than higher ones and in Autumn lower areas are watered for longer. Where a bisse is required to irrigate both vines and meadows, the former are watered by day and the latter by night.

Historically, watering was not done on Sundays or holidays but dispensation has become more widespread. In order not to waste the water on those days, many bisse systems include ponds for storage. Those on the plateaus of Savièse, Ayent and Lens are of glacial origin. At Savièse there are six of them, known as the Gouilles de Savièse, and they serve as "déversoirs" for the six branches of the Bisse de Savièse. The water in the ponds is not available to anyone who has no right to the water of the Bisse; it is accumulated until there is enough for a day's watering.

The Ayentots, faced with the problem of what to do with Sunday's water, allocated that from the Grand Bisse to Grimisuat under the terms of an agreement made in 1464, in exchange for Grimisuat's sharing in the work of the upkeep of the bisse, and left it to its recipients there to obtain dispensation from the bishop to water on Sunday afternoons.

Later, the Sunday water from Bitailla, on the advice of the good Curé Cordel, was put into four ponds, at Blignoud, Botyre, Saxonna and Fleives (fig. 28). The Grand Bisse also fills two ponds on the territory of Arbaz. There are five ponds in the commune of Lens.

From early times, storage ponds were made in suitable locations by building small dams, e.g. in 1623 the Illsee was created to store the snow-melt. In the original dam the stones were stuck together with pine resin, which lasted until the Turtmannwerk power plant came into possession of the dam and replaced it with a concrete one.

To complete this account of irrigation in the Valais I should mention the "meunières" (mill-leats, but if ever used as such, certainly not any longer) which are mainly found near Martigny and in the stretch of the Rhone valley between there and the Eastern end of the Lake of Geneva. I have also heard the word used in connection with channels near Conthey. They are man-made watercourses but differ from bisses in that their prises d'eau are not high up in the side valleys but at points where the torrents reach the main valley.

Accordingly, they do not suffer from the difficulties, dangers and complications of bisses; maintenance is, of course, much easier, but they are equally important for the irrigation of the fields in the valley bottom. Many of these fields were developed in areas which were marshes until the disciplining of the Rhone (see page 19); some of the meunières are, therefore, relatively modern, but at least one, which takes its water from the Drance,is known to have been operating in 1723. Because, at this altitude, water is rarely scarce, regulations are much less precise and strict than those for the bisses, but the system of operating is similar. Flow from the main channels into smaller branches is controlled by sluices and the final divisions are made by cutting temporary holes in the banks. The tools used – the torniou (plaque d'arrosage, tranchant) and the délabre (pioche) – are identical with those used on bisses.

Fig 1: Bisse d'Ayent at Croix, as it looked just after restoration work started – so far only the wire rope had been fixed.

Fig 2: *Bisse d'Ayent at Croix: Firmin Morard and Armand Dussex at work near the "window". (Photo: Firmin Morard)*

Fig 3: *Bisse d'Ayent at Croix: first new section completed. The top of the date, 1991, cut into the second boutset in the traditional manner, can just be seen.*

Fig 4: Bisse le Lentine. Prise d'eau from the Sionne with, from left to right, over-flow back into the river, sluice (closed) to direct the water back into the Sionne when the bisse is to be emptied, grille to protect the bisse, sluice (open) controlling the input of water into the bisse which, for the first few metres, is in a pipe under the walk-way.

Fig 5: Bisse de Lentine crossing a masonry bridge over a torrent by means of a metal channel, with sluice to allow the bisse to be discharged into the torrent in the event of an emergency further down the bisse.

Fig 6: Niwärch Suon. Modern metal Wasserschlegel (water-hammer). Compare with Fig. 25

Fig 7: Niwärch Suon. Unirrigated land at the top, an irrigated meadow on the near side of the suon, with wooden sluice gate and Wässerhaue (watering-pick)

Fig 8: Grand Bisse de Lens. On the 3km. stretch which, having been replaced by the Mont du Châtelard tunnel, is maintained as a tourist attraction. Overflows and breakages used frequently to cause damage to the village of Chelin below.

Chapter 7

Number, Lengths and Names of Bisses

The total number, lengths and names of bisses are subjects on which it is impossible to be precise, mainly because, as must be clear from previous chapters, they are all very individual, entirely locally-controlled and not effectively centrally recorded. The lists which I have consulted and which, unfortunately, do not all contain exactly comparable information give the following figures:

    a) Leopold Blotnitzki, (1871) followed by H. Hopfner (1898)
        Total number: 116
        Total length: about 2000 km.
    b) Fritz Rauchenstein, (1908) modified
        Total number: 207
        Total length: 1397.4 km.
    c) Theo Schnyder, (1926) quoting only totals with no breakdown
        Total number: 300
        Total length: 2000 km.
    d) Service de l'Aménagement du Territoire de l'Etat du Valais (1993)
        Total number: 190
        Total length: 731.4 km.
    e) Emmanuel Reynard (1994)
        Total number;376
        Total length: 1748.5 km.

This surprisingly high total number was reached by meticulous study of all available inventories and the 1:25,000 maps. The result includes some with no individual names and some which not only are out of use but which the previous compilers did not record, presumably because they never found or heard about them. It also includes a number of very short water courses which other compilers may have regarded as not qualifying for the designation of bisse; it does not conflict with the indications in preceding chapters that the number in use has been steadily falling.

Of the highest total of all bisses quoted above, 376, only 37 (most of them among the oldest) originated at over 2,000 metres; of these, 16 are still in use, four recorded as out of action and the balance apparently lost.

Figures for the lengths of some bisses given by different writers vary by up to several kilometres; in any event over a dozen are more than 20 km. and the abandoned Bisse de Saxon, the longest of all, was 32 km. Further discrepancies seem to be due to some of the bisse systems being sometimes referred to by one or more general names, e.g. Bisse de Savièse, Torrent Neuf or Bisse de St.Marguerite, and sometimes by the names of some or all of their individual components. The extent of this confusion can be gauged by e.g. the existence of 62 names for the whole and the sections of the Bisse de Savièse. Many are descriptions of the locality, e.g. Liviou (place where water is drawn off a torrent); Escortiou (tunnel through scree); Fô (rock which looks like an oven); Pari blantze (white rock face); Fontannaz du Boutzé (small spring which was struck when making a hole for a support beam); Zour de Frano (ash wood); la Zemma (twins – two big springs side by side); Zena de l'Ours (place where a bear was killed); Barma nire (black wall). In the Swiss-German-speaking area the same applies, e.g. on the Niwärch at Ausserberg, Z'Leid Eggi (mourning corner); Todflühe (death rock-wall); Gamschsprung (chamois jump); Nidler (place where the girls from Raaft brought fresh cream ("Nidla") to the men working on a "chänil" (hollowed-out tree trunk)); Äuchublattjigrabo (place from where those girls from Raaft fetched slate slabs on which to put their butter-tubs).

In both languages, many bisses built in areas where there were – or had been – earlier ones were called "new" (neuf, niva) and are still so called, several hundred years later (Torrent Neuf, Niwärch). The names of some bisses indicate clearly their source, e.g. Bisse des Audannes (Ohannes); others their destination, e.g. Bisse de Sion; others indicate only to those with detailed local knowledge, e.g. Bisse d'Azerin has its source at Losentse and irrigates at Chamoson and Bisse Vieux or d'En Haut has its source on the Printse and irrigates at Nendaz. Many bisses in Haut Valais have names ending -erin, a feminine word ending, often shortened further to -eri, which some regard as an abbreviation for suon, e.g. Ebibergerin for Ebibergersuon, Riederi for Riedersuon, but it seems to me more likely that these were originally affectionate female personifications of the bisses indicating origin, e.g. Augstborderin, the bisse – or girl, whichever you will – from Augstbord, Milleri, the same from Milibach.

Since the main object of this book is to awaken interest in bisses, I hope that, after reading it, people will want to go and look at and walk along at least some of these extraordinary achievements. Emmanuel Reynard has very kindly rearranged for me the inventory compiled and made public in 1993 by the Service de l'Aménagement du Territoire de l'État du Valais, so that one can see at a glance what is working (167 in 1994) and what is not (23) – and plan one's itineraries accordingly. You will find this in the Appendix.

Fig 32: Grand Bisse d'Ayent. Reconstructed section at Croix, seen from the opposite side of the gorge. The new viewing platform can be seen protruding from the enlarged "window"; the entrance to the 1831 tunnel is just to the right of the reconstructed section.

Fig 33: Grand Bisse d'Ayent. Reconstructed section at Croix. The tunnel entrance is hidden by the dark-coloured rock on the right.

CHAPTER 8

Recent Developments and Future

The earliest important change in the construction and maintenance of bisses was the making of tunnels – which, except for very short lengths, presupposed the use of explosives. Many of the earliest tunnels, of necessity short, had head-room of only about a metre, so the tunnellers had to work lying down. Among the earliest must have been the five on the Bisse de Savièse, the building of which began in 1430. Crawling through them in the early 1930s was described by a Canadian, Douglas Stevenson, in the 1955 Canadian Geographical Journal. The 7.7 km. Bisse de Clavoz, built in 1453 and of the greatest importance to the vineyards of Sion, was gradually improved with tunnels in the course of the centuries and now has a total of about 1,200 m. of them. In 1831 on the Grand Bisse d'Ayent, the Torrent-Croix Tunnel, almost 100 m. long, was bored, to replace a wooden cantilevered channel across a vertical rock face; it is recorded that 1,000 lbs. of black powder were used. The tunnelling went slightly off course at one point and broke through the cliff face from behind, thus producing a window which has been very helpful ever since to all those using the walkway through the tunnel. It also produced an argument with the entrepreneur, who thenceforth defenestrated the spoil instead of carrying it back to the end of the tunnel and the consortage argued that, for this economy of effort, his price should be reduced. The window was enlarged in 1991 when, in a rather belated realisation of the heritage and touristic interest in bisses, part of the wooden channel was restored using traditional methods and materials supplemented by modern safety precautions, such as a wire rope anchored with rock-bolts to which those working could attach themselves (figs *1*, *2* and *3*, 32 and 33). A viewing platform was built from which one gains a splendid impression of that channel and of the difficulties and dangers which its original construction and maintenance must have presented. At the Northern end of the tunnel – i.e. the Northern end of the original wooden channel – there used to be a primitive crucifix, as was the practice in many dangerous places along

Fig 34: Grand Bisse d'Ayent. Primitive crucifix, formerly at the Northern end of the Torrent Croix tunnel, stolen in the 1980s and replaced by a modern one. The wording reads: "Mortels! Respectez ces lieux et priez pour les âmes souffrantes 3p. 3a les agonisants 300 jours d'indulgences. Fardel, curé, 1863".

bisses (fig. 34). It was stolen in the 1980s and replaced by a modern one.

In August 1914, Ausserberg, which has been plagued with chronic water-shortages since time immemorial, suffered the cruellest blow possible: all three water courses from the Baltschiedertal, the Niwärch, Mittla and Undra suonen, were carried away by an immense rock-fall. A series of very courageous attempts at repair failed because rock-falls continued and for a time survival depended on the small spring at the Bern - Lötschberg - Simplon railway station. To add to these crises, all the men liable for military service had been called up to man the frontiers because of the outbreak of World War I. Eventually their release from military service was negotiated and, after two months of work day and night, the three suonen were flowing again, through tunnels 130 m., 80 m. and 80m. long respectively. The first of these bypassed the notoriously dangerous passage known as Z'leid Eggi (fig. 12). However, it became clear that, from a long term point of view, this

Fig 35: Niwärch. Distributor at the Southern end of the 1972 tunnel, which brings the water of the Niwärch and Mittla suonen through instead of round the mountain.

Fig 36: Niwärch. Tunnel entrance at the distributor ("Strahlen" means – only in Switzerland – seeking and extracting crystals from rock)

87

Fig 37: Niwärch. looking into the tunnel from the Southern end. Niwärch and Mittla flow together on the left.

was not enough and, after playing for a time with the idea of pumping water right up from the Rhone, a scheme was drawn up in 1967/68 to bring water from the plentiful springs in the Baltschiedertal, hitherto carried by Niwärch and Mittla, through a new channel over 5 km. long, of which 1.6 km. are in tunnel. Work began in 1969, took three years and cost 4.2 million francs, of which the Federation paid 35% and the Canton 33%. This solved the problem once and for all (figs. 35, 36 and 37). Very surprisingly, in spite of all the mishaps which recurred on Niwärch's precipitous course along the Western rock-face of the Baltschiedertal, there is no record of any fatality there in the 600 years of its existence until 1960, when the Wasserhüter fell to his death.

In 1919 the 2,500 m. Gebidem tunnel was built to replace the courses of

Fig 38: Gasperisuon can be seen, near the middle of the photograph, crossing a gorge on the Eastern side of the Baltschiedertal.

the two avalanche-prone and dangerous Niva Suonen above Visperterminen; it increased the throughput of water by a factor of 4 and brought it out 200 m. higher up the mountain, thus enabling a lot more land to be irrigated. In Spring 1935 the 4,800 m. tunnel of Prabé was opened on the Bisse de Savièse to replace the dangerous long wooden channel across the rock face of the Paroi des Branlires (see page 35).

In 1945, after seven years' work, the 3 km. long Riederhorn tunnel was opened, bringing water from near the snout of the Grosser Aletschgletscher to a point high above Ried, replacing three dangerous parallel suonen, including the Oberriederi, which were increasingly difficult to maintain and had claimed the lives of, among others, two successive Wasservögte. Catherine Bürcher-Cathrein's novel "Der Letzte Sander von Oberried" draws on one episode of the long history of six desperate attempts, all using, at least in part, different routes including the Massa gorge, to bring water to Ried-Mörel. The whole story reveals most courageous efforts by a determined community being frustrated by the arguments, obstinacy and delaying tactics of a self-interested few. The third of the routes, which ran for about 10 kms. round the whole Riederhorn massif, was dreadfully dangerous and its maintenance was reckoned to kill the best man working on it each year.

Fig 39: Bisse de Sion. Attractive section of the Bisse de Sion near Le Dailley before it was put into a pipe for the Anzère drinking-water supply.

During 1947, after much trouble, a 2,000 m. tunnel through Mont-la-Chaux was opened on the Bisse de Ro, but unfortunately proved useless and became an access track to the alpages. In 1949 substantial improvements were made to the Augstborderin suon: 2.5 km. out of its length of 11 are now in tunnel and most of the rest is in concrete pipes. In 1984 the 845 m. tunnel through Mont du Châtelard (which incorporates a waste water channel under the bed of the bisse) was opened on the 13.8 km. Grand Bisse de Lens; this shortened that watercourse by 3 km., avoided a 25 - 30% loss of water by seepage which used to occur over that stretch and put an end to a recurring danger to the village of Chelin and properties below (fig. *8*).

The problem of seepage has plagued the builders and operators of bisses throughout their history. Cement in various forms has long been used to try to seal the beds of bisses and, in particularly bad areas of bog, broken rock, scree and alluvial soil, wooden, then galvanised iron and, later, aluminium trough sections were laid, usually above ground level (see figs. 17 and 18). Iron and concrete pipes followed and then a revolution in this field was brought about by the introduction of plastic pipes, delivered when necessary by helicopter to the locations where they are required. One tunnel, which was never made, was part of a 1918 project to improve the Bisse du Levron by avoiding rotten schistose rock at Pierre Avoua at a height of 1500 m. The project was abandoned and so was the bisse in 1923. Several revised projects in that area also failed but in March 1946 there was a development which changed everything and, in retrospect, demonstrated the possibility of associating the old world of bisses with the new world of hydro-electric schemes. An engineer named Albert Maret was at that time designing the Mauvoisin scheme (Forces Motrices du Mauvoisin) and, in response to an enquiry from the President of the Commission of the moribund

Fig 40: Grand Bisse d'Ayent. Disused section below Les Rousses with a length of banquette still intact.

Fig 41: Grand Bisse d'Ayent. Just round the corner from fig. 40, where bisse and banquette have disappeared, about 15 years later.

Bisse du Levron, he wrote: "our pressure tunnel will come out East of Pierre-Avoua at about 1,750 m. . . . we will put the water of the bisse through our tunnel". In fact that was by no means the end of the story; seemingly endless discussion and arguments about quantities and rights of all parties, including the communes of Le Levron, Bagnes and Vollèges, went on for some time, but by 1957 Mauvoisin was working, two tunnels, one 3,600 m. long under the Bec-des-Roxes, and the other 700 m. long under Pierre Avoua had been bored and water from Louvie and La Chaux had been brought to the Col du Lin, shared out and delivered to Le Levron, Vollèges, Vens and Chemin – and the quarrels of centuries were at an end.

Another modern development, altering fundamentally the use of a bisse by incorporating it into the meeting of present-day requirements, involves the Bise de Sion. This was built in 1905, from Rawyl to Sion, to supplement the waters from springs called "La Fille" and "La Fillette" near Arbaz to meet the city's, its vineyards' and its orchards' increased requirements. Its entire length, 14 km. is on Ayent territory and its building and use were only made possible by a concession to Sion by Ayent. However, until 1970 Ayent had no benefit from it. In that year an arrangement was made whereby water from the Lac de Zeuzier (behind the Rawyl Barrage) is drawn off from the Electricité de la Lienne

SA's tunnel to the underground generating plant at the Six du Samarin and pumped up about 100 m. into a 70 cu. m. reservoir serving the Bisse de Sion. From there the water flows through a pipe, laid in the bed of that bisse and capable of carrying 450 litres per second for 3.5 km., to a filtration plant at Audey. The whole of the original course remains and the bisse can be operated above and below the piped section; no water is available from the springs at Zeuzier during the winter but since the barrage was built in 1958 water can be drawn from the lake. The piped section, under the terms of an agreement made on 5 June 1970 between the Communes of Sion and Ayent, is their joint property, in the proportion of $5/9$ for Sion and $4/9$ for Ayent (250 litres per second for Sion irrespective or origin and destination of the water and 200 litres per second for Ayent). The running expenses of the bisse are divided in the same proportion. At Audey there is a 460 cu. m. reservoir with a bypass for the necessary distribution, including that during periods of watering. After treatment, Ayent's share is distributed, mainly to Anzère, through the mains of the commune of Ayent. That same point on the Electricité de la Lienne's tunnel is now also the source of the Grand Bisse d'Ayent which, from there upstream to its original prise d'eau on the Lienne, is derelict (figs. 40 and 41), but downstream, fed from the tunnel, is fully operational.

The availability of new equipment and techniques towards the end of the 19th century sometimes brought with it a touch of arrogance; some lessons had to be learned the hard way. One example was the sad story of the tunnel on the Bisse du Ro, which has been mentioned above, and another was the unfortunate attempt by the Commune of Vollèges in the Bagnes valley to increase its water supply, which depended on two bisses, so that the alluvial cone near the village of Cries could be irrigated. Rather than build a third long and expensive bisse, it was decided to instal a turbine-driven pump on the floor of the Drance gorge, a short distance upstream from the confluence of the Merdenson torrent, to raise 50 litres per second 145 m. The scheme was a disaster because no-one had realised that, in summer, the water of the Drance is heavily charged with a very abrasive fine sand which speedily wrecked the machinery.

As mentioned before, the number of bisses in use has declined in the 20th century – and not just because some have been destroyed by natural causes. Replacement by pipes and changes in the requirements

Fig 42: Bisses crossing without mixing near Étang Long (Arbaz).

Fig 43: Bisses crossing without mixing near Pro Catroué (Arbaz).

for irrigation have all played their part. The need for maximum cattle and crop production has decreased because very few mountain farmers are now entirely dependent on the fruits of the soil and this has resulted in less watering of the fields. On the other hand, viticulture's requirements grow no smaller. The overall need for water has increased with increasing population of the mountain areas and the expansion of the tourist trade; the sources of water are as important as ever, the avoidance of wastage is much more efficient and water-rights are as fiercely defended as ever, so that the builders of hydro-electric installations have to take them into consideration in their planning as, for example, the Force Motrices du Mauvoisin, the Grande Dixence S.A. and the Electricité de la Lienne S.A. all recognised in good time; in the Ayent area the Electricité de la Lienne pay the communes and the consortage for the water it takes and these funds go towards the upkeep of the bisses. The bisses continue to have first call on the water and the practice of keeping the water of different bisses separate from each other also continues (figs.41 and 42), except in emergencies or when the construction of a tunnel makes it sensible to combine them (e.g. Niwärch and Mittla at Ausserberg (figs. 35, 36 and 37)). Some anomalies inevitably occur, especially where water-rights and land are sold separately. At the first International Colloquium on Bisses in Sion in September 1994 someone asked how he could get rid of an unwanted collection of water-rights. The answer was good Valaisan philosophy: "calculate their value and drink it!".

The Bisse de la Tsandra, built by a consortage in 1400 to bring water from the Morge to the villages of Daillon, Erde and Aven was jointly owned by them and the 1,234 water-rights (each divisible into halves or thirds) were divided equally between them. It suffered throughout its history from seepage of about $1/3$ of the water put in at the prise d'eau and from recurring costly damage to property below it. Troubles came to a head with the drought in 1858, when fields and vines suffered very badly. In that year, also, the village of Premploz asked to join the consortage and suggested that the bisse should be enlarged at its, Premploz's, expense. The other three villages refused; they proposed to enlarge the bisse to meet their own needs. Faced with this refusal, Premploz turned to the Conseil d'État, which sent an engineer named Venetz to inspect the situation on the ground and report. His first report said that "the Commune of Conthey should investigate all means of getting enough water for the improvement of the whole area . . . La

Fig 44: Bisse de la Tsandra. The water in the photograph is purely decorative; the bisse is in a pipe under the walk-way.

Tsandra is badly built and worse maintained." Venetz's second report, three weeks later, proposed that the Commune should take over the running of the bisse. The Commune refused and deadlock ensued. On 23rd December 1858 the Conseil d'Etat forbade any enlargement of the bisse until the matter had been sorted out. Furthermore, they said, the water from the Morge did not belong to the three villages and, to cap everything, the Consortage had not distributed the water-rights in accordance with the law. In May 1869 the Conseil d'État at last gave its ruling: the bisse should be enlarged to deliver double the quantity of its flow; the work should be carried out by Premploz; the water-rights should be divided into four equal parts. Two years later these works were completed and, the year after that, Mouzerin (Daillon) demanded that the bisse should be further enlarged by $1/5$th for its use. Aven, Erde and Premploz opposed but were over-ruled by the Conseil d'État – and so things continued until 1977, but the years 1932 to 1962 saw a continuing dispute between the Consortage du Bisse de La Tsandra and the Communes of Savièse and Sion.

At this time the Consorts decided to rebuild the bisse because losses of water were getting worse, the costs of repairs were getting higher and no insurer would cover the risk of damage. Simultaneously, the commune was considering the provision of a separate drinking-water network (hitherto the bisse had provided both irrigation and drinking water). It was also pointed out that, in addition to the loss of $1/3$ of the water through the bottom of the bisse, there were further substantial losses inherent in the "torrents de décharge" (the many minor water courses which take the water from the bisse to the individual areas to be watered) and also in the method of irrigating fields by "ruissellement", which uses twice as much water as "aspersion". In May 1982 the Canton agreed to subsidise the reconstruction of the bisse and plans for that operation were delivered to the Service Cantonal des Améliorisations Foncières. The project was divided into seven sections and included putting the bisse into a pipe for its whole length, remotely-controlled regulation, measurement and pressure-release to ensure that water flows into the distribution system only when required, thus allowing substantial economies, a 200 m. tunnel through the Roc de Mayentzet, two new sand-traps, a drivable track along the whole length of the bisse and, eventually, conversion to overhead sprinkling of 95% of all watering. The completion of the last section was reported in the Municipalité de Conthey's Bulletin d'Information of July 1993.

While all this was going on, a radical sorting-out of anomalies in ownership was effected. The Bisse de La Tsandra was built as a consortage to irrigate fields rather than vines and the water-rights were not attached to the land. As individual pieces of land were sold for either planting with vines or building, the situation was reached in which an increasing number of individuals had water-rights and no land (see page 95). By 1985 this had become rather absurd and, so, as a first step, the system of distribution was changed to a square metre basis and then in 1993 the Commune bought out the Consortage. The money for this was raised by imposing a tax upon the Commune – and supplemented by an action described in the local press of July 1993 as follows: "in March 1990 an admirer of nature, impressed by the work carried out to maintain the bisse, even underground, paid into the private account of the President of the Consortage the sum of F 50,000. In accordance with the wishes of the beneficiary, this sum will be transferred by his family to the Commune as soon as the Consortage is dissolved and it is to be used to compensate the owners of water-rights . . . "

It is gratifying that, in these times of change there is a widespread, even if not ubiquitous, acceptance of the need for it and a readiness to make use of and modify the resources already available. Apart from that last most modern example, others readily spring to mind including:

> a) 1914-1917 and 1980: modification of the regulations dating from 1448 and 1698 of the Grand Bisse de Lens to bring them into line with modern requirements;
>
> b) 1945-1953: concessions made by the riparian communes of the Lienne (Ayent, Icogne, St. Léonard, Sion) to enable the building of the Zeuzier dam;
>
> c) 1970-1971: formal change to joint ownership (Ayent and Sion) of the Bisse de Sion to provide water for Anzère.

On the other hand, if bisses are not kept working for their original purposes, (e.g. la Tsandra), maintained as examples of a fascinating heritage (e.g. Niwärch) or modified to meet modern requirements (e.g. Sion), avalanches and rock-falls see to it that parts of them soon disintegrate, as the most impressive section of the Bisse de Savièse has done and the disused upper stretch of the Bisse d'Ayent is rapidly doing; about 15 years separate the two photographs shown in figs. 40 and 41. However, after a period in which bisses were regarded as old-fashioned and of little importance, I believe that the last few years have seen a revival of interest in them, brought about by a recognition of their immense importance in the history of the Canton.

ENVOI

Until about 60 years ago, all those who worked on or in bisses took it for granted that they would always have wet feet. A thoughtful Anniviard, with vivid childhood memories of helping his father irrigate the family's lands, reckons that the most important development in connection with bisses in his life-time was the invention of rubber boots . . .

CHAPTER 9

21 Suggested Excursions

Essential Equipment:
1:25,000 map sheets, small torch for tunnel sections, one or two cars and/or post bus and train timetables.

To re-state the obvious, bisses follow the contours at gentle gradients. This both enables them to be readily distinguished on the map from torrents, which cross the contours much more frequently, and means that, in general, they offer gently-graded walks which can be enjoyed by the youngest and the oldest members of a party.

From the number of bisses already mentioned in the preceding chapters – and the many other ones listed in the Appendix – the tentative investigator might well wonder where and how to begin. I have, accordingly, selected below 21 excursions as starters. Except as noted, they can all be walked in either direction; numbers 14 and 18 and part of number 11 should not be tried by anyone subject to vertigo. Driving one's car to as near as possible to one end or the other of a bisse often necessitates walking it in both directions – which is fine provided that up to ten hours walking is acceptable. Alternatives are to use two cars by positioning one at or near the distant end before starting, or to use public transport. As will be seen from the itineraries below, public transport is generally available, but a caveat is necessary: on some post bus routes the services, reasonably enough, are few and far between and some cable car operating times are very limited, especially out of season. Local enquiries are essential. The following suggestions are not intended to be an exhaustive tourist guide but to get you moving in the right directions.

1. BISSE DE TRIENT
    Map sheets 1324 and 1325.
    Post bus Martigny to Col de la Forclaz, 1,526 m. – pick up the bisse there; follow it South through Ban du For, Le Tsaperon and L'Argny to

the prise d'eau at the glacier, then back to cross the Trient torrent at point 1,550 m. and North on the West side of river through Les Petoudes and Le Peuty to Trient village. Return by post bus to Martigny.

> COMMENT – A straightforward half-day's walk on both sides of the Trient valley.

2. ANCIEN BISSE DU LEVRON
   Map sheets 1325 and 1326

   Train Martigny to Le Châble. Cable car to Les Attlas, 2,733 m. Follow a well-contoured track South-East to the Mont Fort Hut, 2,457 m. Turn due West and follow course of the bisse to Les Ruinettes, 2,193 m., thence North and North East to Les Planards, 1,930 m. and return to Verbier, 1,500 m., by any one of a number of paths. Return by cable car to Le Châble and thence by train to Martigny.

   > COMMENT – A fairly strenuous day's walking which demonstrates some of the troubles experienced by the early bisse-builders in trying to bring regular water-supplies through a very difficult area.

3. ANCIEN BISSE DE SAXON
   Map sheets 1305, 1306, 1325 and 1326.

   Post bus Martigny to Col des Planches, 1,411 m. Follow the bisse up its course North East through Pas du Lin, Boveresse, 1,499 m., and La Tsoumaz to Mayens-de-Riddes, 1,700 m., or beyond. Return by post bus Mayens-de-Riddes to Riddes or cable car Isérables to Riddes.

   > COMMENT – This bisse runs in a broad arc South of Riddes and Isérables; the length of the walk depends on the point at which one leaves the bisse to turn North towards these villages. It should be borne in mind that the total length of the bisse is 32 km.

4. BISSES VIEUX AND DU MILIEU, NENDAZ
   Map sheet 1306.

   Post bus Sion to Haute-Nendaz cable car station: walk West up a metalled road to Bisse Vieux, 1,444 m.: turn left and follow up the course of the bisse through Eterpay, La Bertouda and L'Antié (in the Printse valley) towards the prise d'eau on the river South of Planchouet. Rather than going to the prise d'eau, take a path downhill to the left about ¾ km. South of L'Antié, cross the Printse at point 1519, turn left along a track through Planchouet, take a path to the left just before La Tsâche torrent to re-cross the Printse below the prise d'eau of the Bisse du Milieu. Follow the bisse downstream through La Biola and Cerisier back to Haute-Nendaz (centre). Return by post bus to Sion.

    COMMENT – an easy half day through woodlands and meadows.

5.   BISSE D'EN BAS (DE DESSOUS) DE NENDAZ AND BISSE DE VEX
    Map sheet 1306

Post bus Sion to Basse Nendaz. Pick up Bisse d'en Bas at 988 m. just West of road to Beuson, follow up its course through Plan Torrent to its prise d'eau on the River Printse. Follow up the course of the river for about 1 km. past Achouet to the prise d'eau of the Bisse de Vex on the right (East) bank of the river just South of Planchouet. Follow down the Bisse de Vex northwards round the spur of Le Bandé, crossing the torrents of Les Rontures, Le Doussin and Ojintse and North and then North East round the next spur to Mayens-de-Sion. Return by post bus to Sion.

    COMMENT – A full day's easy expedition; this can be shortened by taking the post bus to or from Planchouet and following one or other of the two bisses. Parts of this route give excellent views North to the Bernese Alps.

6.   ANCIEN BISSE DE CHERVÉ (CHERVAIX)
    Map sheets 1306 and 1326.

Post bus Sion to Thyon 2,000. Follow the line of the bisse North towards La Matse, round the North end of Crête de Thyon and South past points 2,136m. and 2,160m. to La Combire 2,209 m., Combartseline 2,204 m., past the top station of a cable-car from Super Nendaz, past Chervé and Tsava, above the East side of Lake Cleuson, to La Gouille, thence steeply down to the head of the Lake, 2,201 m., along the West side of the Lake to the dam and down the valley to Super Nendaz, 1,733 m. Return by post bus to Sion.

    COMMENT – a fairly strenuous day with splendid views.

7. BISSE DE LA TSANDRA
    Map sheets 1306 and 1286.

Post bus Sion – Grand-Dzou, 1,437 m. (on the bus route to Barrage de Sanetsch) – first go about 500 m. North along a track parallel to the road (and in fact covering the bisse which is in a pipe) to see the prise d'eau on the Morge, then retrace your steps and carry on along the course of the bisse (in pipe all the way) through Plan Cernet, Mayens-de-My, Mayens-de-Conthey to Aven, 931 m. Return by post bus to Sion.

> COMMENT – An easy day's walk through varied countryside including forests at one end and vines at the other. Looking across the gorge of the Morge, much of the old course of the Bisse de Savièse can be seen. The Bisse de la Tsandra is an interesting example of the complete modernisation of an ancient water-course.

8. BISSES ON THE SAVIÈSE PLATEAU
    Map sheets 1306 and 1286.

The opening of the Prabé tunnel in 1935 revolutionised the water supply of the whole plateau (see page 35, 47 & 89) and moving about on it now or, indeed, looking at the 1:25,000 map (especially the older edition of 1965 which shows buried water courses which have been omitted on later ones), one gets the impression that there is water everywhere, in streams, bisses and ponds. I have not succeeded in working out a satisfactory round trip, but offer the following two ideas, which can be combined if desired:

> a) Post bus Sion to St. Germain. Walk North East to Prinsières (South of Monteiller) and pick up Bisse de Dejour (or Dijore) on the left of the path and follow it through fields and forest past Pra Bacon, 1,053 m., to the Drahin torrent, which it now crosses in a rather amateurishly-installed suspended pipe. If you look upstream to your left, you will see a rock face at the foot of which are the remains of the cantilevered wooden channel which previously carried the bisse and which has collapsed during the time for which I have known the area. Go back to Pra Bacon and turn right, up either of two paths which climb quite steeply through the forest to the Torrent Neuf. Then either:

    i) follow it to the right up to the Prabé tunnel at La Pétouse, just above Mayens-de-la-Dzou, 1,343 m. and return by post bus from there to Sion, or
    ii) turn left along the bisse as in b) below;
b) Post bus to Mayens-de-la-Dzou. Walk the short distance to La Pétouse and there follow the Torrent Neuf (buried for the first 300 m.) to the East down past the point where you would have joined it in a) above but, instead of descending to Pra Bacon, carry on along the bisse through the forest to a point just beyod Tramillau where the Torrent Neuf is joined by another bisse coming from the West. Leave the Torrent Neuf (which continues downhill towards some ponds and St. Germain) and follow this other bisse which is, in fact, the lower end of the old course of the Bisse de Savièse and rises gently, past Ninda, to the point where it emerges from a large pipe (which brings water by a direct route, buried all the way, from La Pétouse). Carry on for 1/2 km., along the old course of the Bisse de Savièse, and you come to Ste. Marguerite's chapel, just beyond which the famous wooden section of the Bisse de Savièse used to begin. Return to St. Germain and post bus back to Sion.

COMMENT – One can easily spend a day looking at the network of channels on the Saviese plateau, while enjoying spectacular views across the Rhone valley to the Valaisan Alps.

9. BITAILLA (BUILT BEFORE 1300)
  Map sheet 1286
  Post bus Sion to Anzère. Walk towards but not as far as the cable car station, disregard the Bisse de Sion and follow signs for Le Go, across a small plateau. At the Western end of that plateau the (unmetalled) road swings to the left, but go straight ahead down a fairly steep path which bears to the right (North) and descends into the Sionne valley to

join another unmetalled road. After about ¹/₂km. take the left fork which continues downwards to the Répartiteur at Les Yvouettes (Dorbon), also known locally as Diviseur de la Diète. Here Bitailla takes its water from the Sionne and runs South, almost parallel to it but keeping well above it. Near point 1344m. it crosses without joining the Bisse de Sion, contributes to a pond and then divides several times ("les 5 eaux") to supply Saxonna, Botyre, Blignoud, Fleives (Lombardon) and Luc (fig. 28). Return by post bus from St. Romain, Blignoud or Botyre to Sion.

> COMMENT – I suggest you do not try to walk this bisse in the reverse direction unless you start only at point 1344 because, as you will see from the map, the bottom end of it is very confusing.
> – An easy half-day with splendid views from Anzère across the Rhone Valley to the Valaisan Alps and up the Sionne valley to the Wildhorn.

10. BISSE DE SION
    Map sheet 1286.

Post bus Sion to Barrage de Zeuzier. Take the path on the West of the lake towards Lourantse if you wish to start at the prise d'eau at the Northern corner of the lake. Then return to the barrage and walk along the road through the tunnels - there is a short gap of "daylight" just before the Southern end of the longer tunnel, where the bisse can be seen between the East side of the road and the rock face inside which it runs in a low tunnel. Pick it up West of the road at Les Rousses, 1,767m. and follow it to Six du Samarin, where it goes into a pipe and stays so past the unstable area of Ravouéné, where it joins the route of the disastrous Bisse des Audannes (see page 30) and passes through the filtration plant at Audey, to reappear only at a point near Les Bochonesses. Follow the bisse through Anzère to Pro Catroué where it crosses but does not join Bitailla and the Bisse d'Arbaz and drops down into the Sionne. Return via point 1207m. to Arbaz and thence by post bus to Sion.

> COMMENT – The course of this bisse is very obvious: even where buried it is clearly sign-posted and the walk-way is well-maintained. A comfortable day's expedition with excellent views of the Valaisan Alps nearly all the way.

11. GRAND BISSE D'AYENT
    Map sheet 1286.
    Post bus Sion to Barrage de Zeuzier. The bisse is derelict to Ehéley but the path is walkable with care. Follow it steeply down to Pro du Sex, below Pra Combère, and cross the St. Romain – Zeuzier road below Ehéley, 1,435 m. After about $^1/_2$ km. you come to the gated entrance of a short tunnel leading to the Samarin pumping and generating station in a cavern excavated in the mountain-side. From here on the bisse is operating and uncovered, except for a scree slope on the side of the gorge of the Torrent Croix, where it is in a pipe, and then the 100 m. Torrent Croix tunnel (see page 85). A section of the wooden channel, replaced by the tunnel in 1831, was rebuilt in 1991 by the Consortage du Grand Bisse as a tourist attraction (figs. *1, 2* and *3,* 32 and 33). Until a few years ago there was an interesting primitive crucifix at the Northern end of the tunnel (fig. 34) but it was stolen and has been replaced by a modern one. Go on above Le Partset, over the Torrent de Forniri, cross the St. Romain-Anzère road at Pertou, 1,372 m. and go through La Tsouma, Pro Catroué and Arbaz to Grimisuat. Return by post bus to Sion.

> COMMENT – A comfortable day's walk with good views across the Rhone valley, but not so consistently good as from no. 10 above because this bisse is in the forest for long stretches.
>
> – Anyone subject to vertigo should leave – or board – the post bus at Ehéley and not attempt the section between there and the Barrage de Zeuzier.

12. BISSE DE LENTINE ET MONT-D'ORGE
    Map sheets 1286 and 1306.
    Post bus Sion to Sionne bridge, 602 m. (on the routes to Arbaz and Anzère) – up the minor road beside the Sionne until you come to the bisse. Turn right along it to see the prise d'eau (fig. *4*) then retrace your steps, cross the road and continue along the bisse (fig. *5*) through Diorly and La Mura, where the bisse turns left and goes down to Lac de Mont d'Orge. It leaves the South East corner of the lake and goes round the hill to Pont-de-la-Morge. Return by post bus Pont-de-la-Morge to Sion.

> COMMENT – An easy half-day, through vineyards, with some short sections of the bisse in pipes.

13. BISSE DE CLAVOZ
    Map sheets 1286 and 1306.
    Post bus Sion to Icogne, 1,026 m., then South through Monteiller, Tsamplan, 868 m., past an old generating station ("Anc. U.E.") on the Lienne. The path is on the right bank. The first 1,000 m. of the bisse are in a tunnel and remnants of an older tunnel can be seen on the right of the path. Apart from one short tunnel, the old walk-way in the Lienne gorge is still used and, although quite vertiginous in parts, it is provided with a barrier throughout. Once out of the gorge, the bisse runs its whole course through vast vineyards, below Signèse and above Molignon, and descends steeply into the Sionne near point 630 m. on the Sion-Champlan road. Return to Sion by post bus or on foot.
    COMMENT – For long stretches the bisse itself and substantial very steep, terraced vineyards above and below it are supported by impressive masonry walls. An easy half day.

14. ANCIEN BISSE DE RO
    Map sheet 1286.
    Post bus Sierre - Crans-Montana, Plans Mayens, 1,620 m. – North to Er de Chermignon, 1770 m. – Praz-du-Taillour, 1,339 m. – Pas de l'Ours – Crans-Montana. A spectacular and in places vertiginous walk high up on the East side of the Lienne gorge. The Bisse de Ro's course, through rocks and forests and past cliffs, led to a troubled history and it was abandoned in about 1949 (see page 28). It was refurbished as a tourist attraction about 30 years ago. Return by post bus to Sierre.
    COMMENT – a strenuous half-day, not suitable for those subject to vertigo. Very impressive (fig. 22).

15. BISSE DE LENS
    Map sheets 1286 and 1287.
    Post bus Sion to Icogne, 1.026 m. – South to point 1057 on the Lens road – pick up bisse on the right hand side and follow it through the forest. You will pass the West end of Le Châtelard tunnel, 845 m. long, bored to cut off a 3 km. run round the mountain which was very costly to maintain because of seepage and also repeatedly did damage to the village of Chelin, lower down (fig. 8). The old walk-way is maintained and provides good views across the Rhone valley. On emerging from its tunnel, the bisse serves Chermignon-Dessous. Return

by post bus Chermignon-Dessous to Sierre.

COMMENT – an easy half day, mostly through woodland.

16. BISSE DE VAREN (RASPILLWASSERLEITUNG)
Map sheet 1287.

Post bus Sierre to Miège. Take the minor road North out of the village until you come to a sign-post pointing right to Bisse de Varen. This path rises steeply to reach the bisse at La Proprija. Turn left for about 500 m. to the Raspille torrent to see the prise d'eau and sand-trap, then back along the bisse through Brand to Varen. Near a sign-post to Varen West, there is what appears to be a naturally ocurring sand-trap. Return by post bus Varen – Sierre.

> COMMENT – after a strenuous start, up to La Proprija, this is an easy walk; the bisse runs mostly through rather stunted forest but there are good views over the Rhone valley. An easy half-day.

17. ERGISCH – TURTMANN WASSERLEITUNG
Map sheets 1287 and 1288.

Cable car Turtmann to Ober-Ems, 1,342 m. – follow the road, then bisse, South through Rooseggini to cross the Turtmänna torrent at point 1,365, then North along the bisse through Mettji, West of Tschärbine to Ergisch, 1,086 m. Return by post bus Ergisch – Turtmann.

> COMMENT – Mostly through forests, on both sides of the Turtmänna valley. An easy half day.

18. NIWÄRCH, AUSSERBERG – EGGERBERG
Map sheet 1288.

Post bus Visp to Ausserberg or train (BLS) Brig to Ausserberg. Go up to the top of the village, where there is a sign post for Niwärch, historical watercourse, and Eggerberg. Take this minor road, cross Lowigrabe near point 1,089 m. and pass Luegji. Near point 1,259 m. the road swings left, but go straight ahead and along the walk-way of Niwärch – along the Baltschiederbach gorge to a point south of Ze Steinu, where you cross the river and follow the Gasparisuon along the other (Eastern) side of the gorge back to Eggen and then Eggerberg. Return by post bus Eggerberg-Visp.

There is an agreeable and very worth-while variant. At point 1,259 m., instead of going straight ahead north into the Baltschieder gorge, follow the road round to the left towards Raaft. After about 200 m. take a track to the left to the end of the tunnel, opened in 1972, which now brings the water from Niwärch and Mittla ("strahlen" in the notice at the tunnel entrance means the excavation of crystals from the rock – an exclusively Swiss meaning of the word) (figs. 35, 36 and 37). Niwärch can then be followed for about 3 1/2 km. to its end, through woods and fields, with many examples of irrigating the latter and a modern metal "Wasserhammer" to signal if the flow stops (fig. 6). Return to Ausserberg and then by train to Brig or post-bus to Visp.

>COMMENT – a spectacular and precipitous long half-day which is not suitable for those subject to vertigo. If you take in the variant as well, you will need a whole day. The variant has only one slightly precipitous passage.

19. AUGSTBORDWASSERLEITUNG (AUGSTBORDERIN)
    Map sheets 1288 and 1308.
    Train Visp (VZ) to Stalden – post bus Stalden to Törbel, 1,497 m. Go North-West towards Moosalp – Chalte Brunne, 2,048 m. – Hoflüe – Läger, 2,108 m. – Embd, 1,358 m. Return by cable car Embd to Kalpetran – train Kalpetran to Visp.

    >COMMENT – a day's quite strenuous walking, much of it over 2,000 m., with splendid views.

20. NESSJERINSUON
    Map sheets 1269 and 1289.
    Post bus Brig to Blatten, cable car Blatten to Belalp, 2,091 m. Go South West to Bäll, 2,010 m., cross the Chelchbach at point 1,968 m. Southwards, dropping gently to the Nessjerisuon on your right to Nessel. Continue South past points 1,862 and 1,740 and so down to Mund, 1,188 m. (the only village in Switzerland where saffron is grown) or Birgisch, 1,093 m. Return by post bus from Mund or Birgisch to Brig.

    >COMMENT – a day's relatively easy walking: if you choose to do this in the opposite direction, however, it is harder work and you should start at Bergisch rather than Mund.

21. FIESCHERTALSUON
    Map sheet 1269.
    Train (FO) Brig to Fiesch – cable car Fiesch to Kühboden, 2,212 m.,

then North to Märjelesee, 2,348 m. – Märjelewang and steeply down into the valley of the Seebach, which flows into the Glingulwasser. Follow the right bank into and along the Fieschertal and on to Fiesch. Return by train to Brig.

COMMENT – A fairly strenuous day's walking, with some magnificent views all round and particularly of the Grosser Aletschgletscher. Doing this walk in the reverse direction is not recommended unless you are in good training; Fiesch is at 1,060 m. and the Märjelesee at 2,300 m.

MAPS

5 examples of typical bisse/suon systems

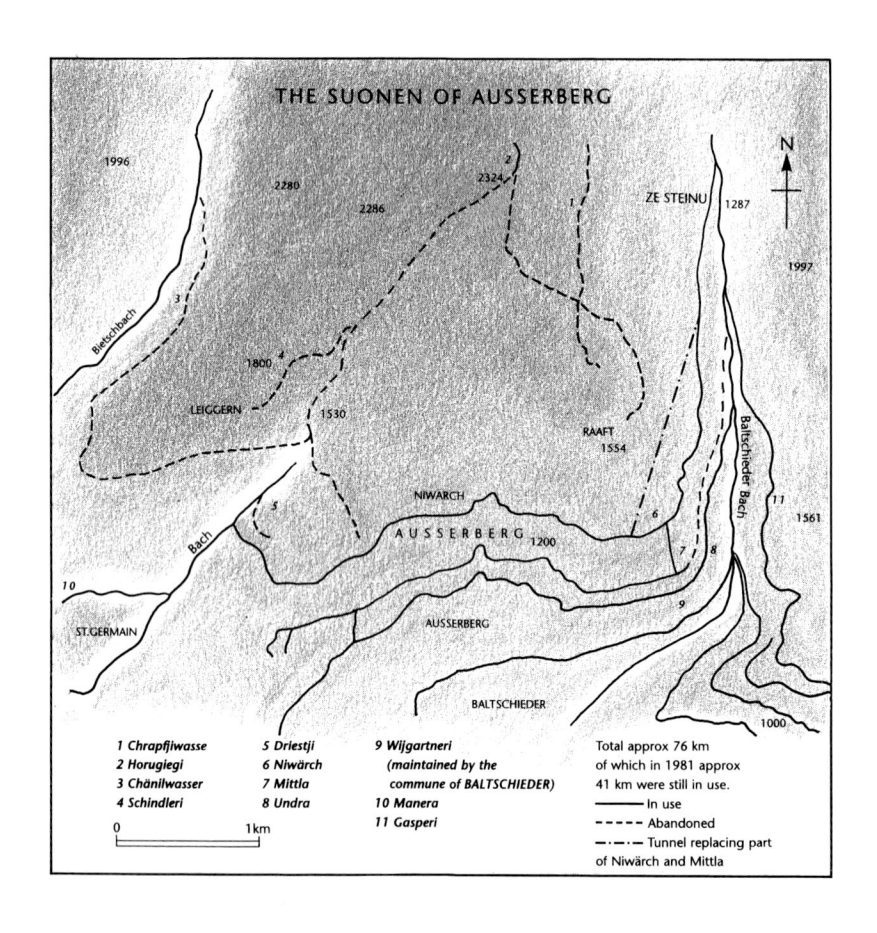

Fig. 45

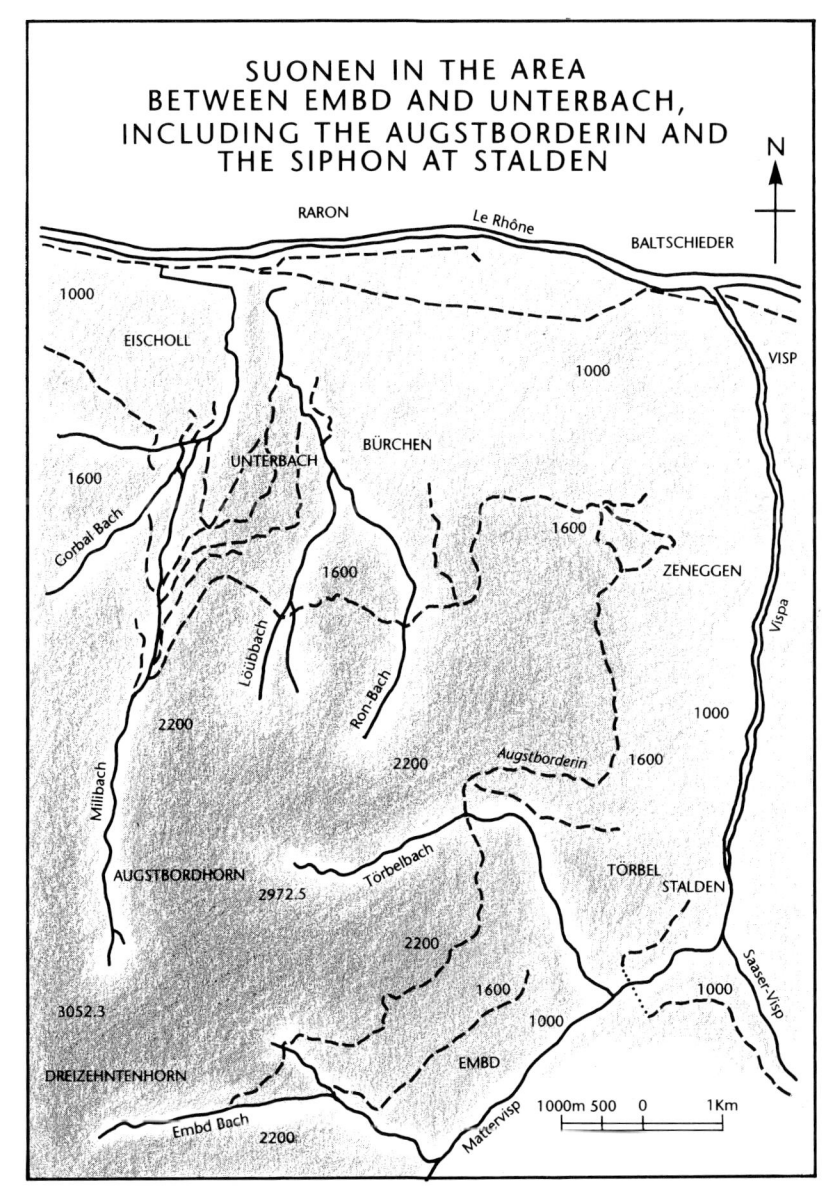

Fig. 46

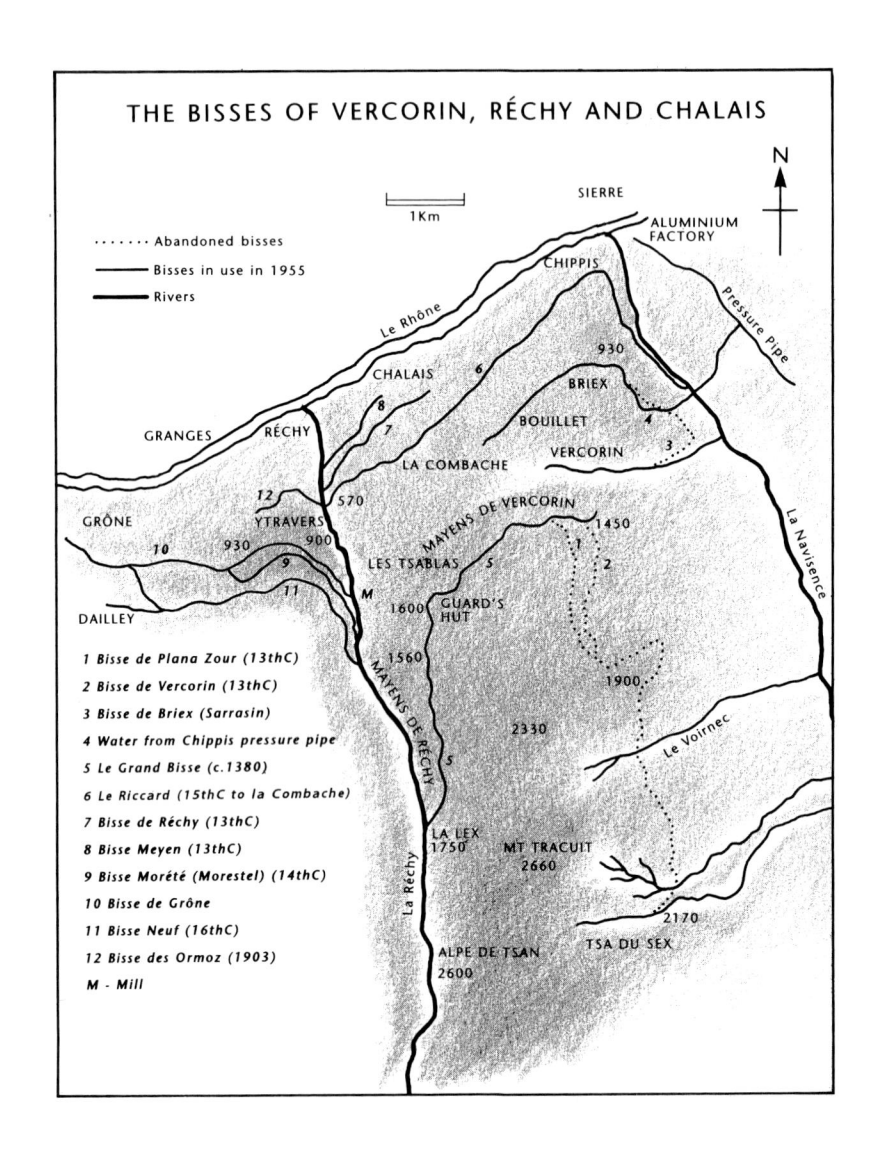

Fig. 47

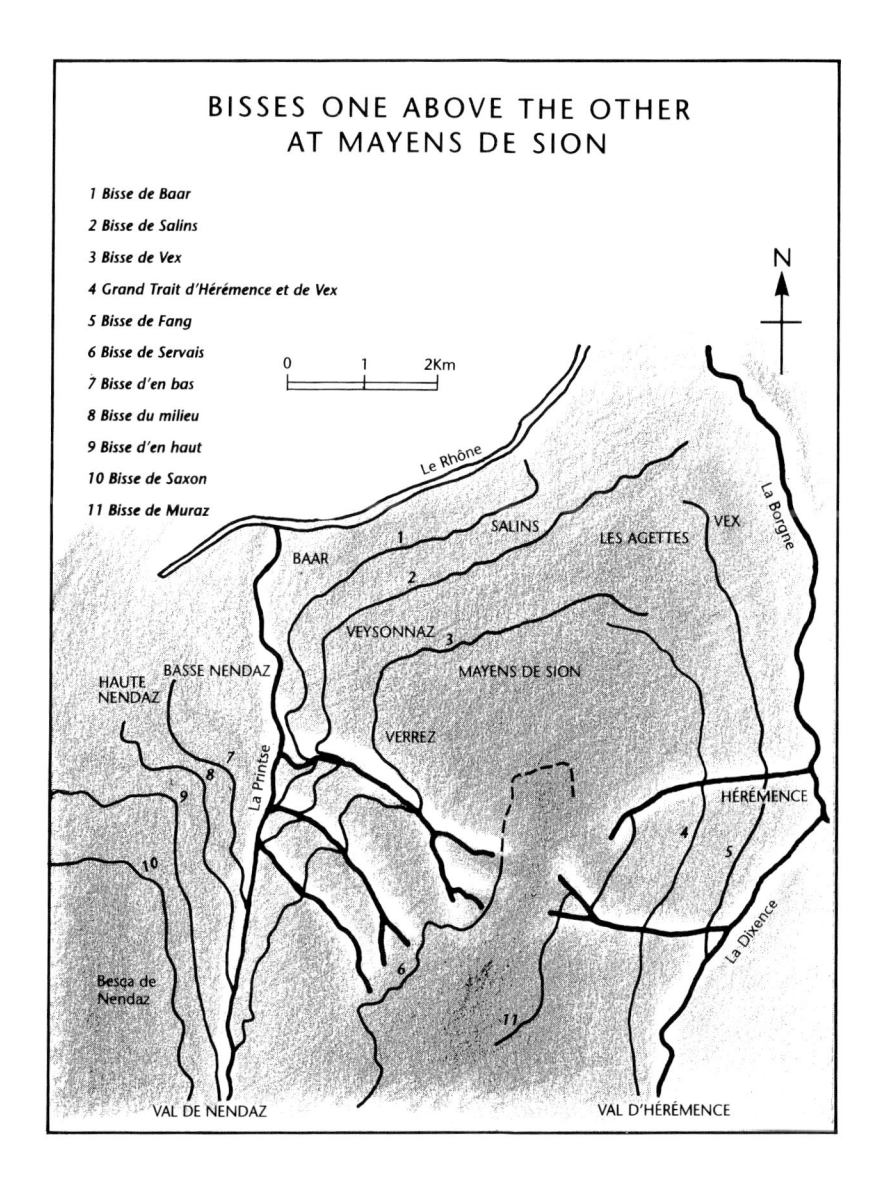

Fig. 48

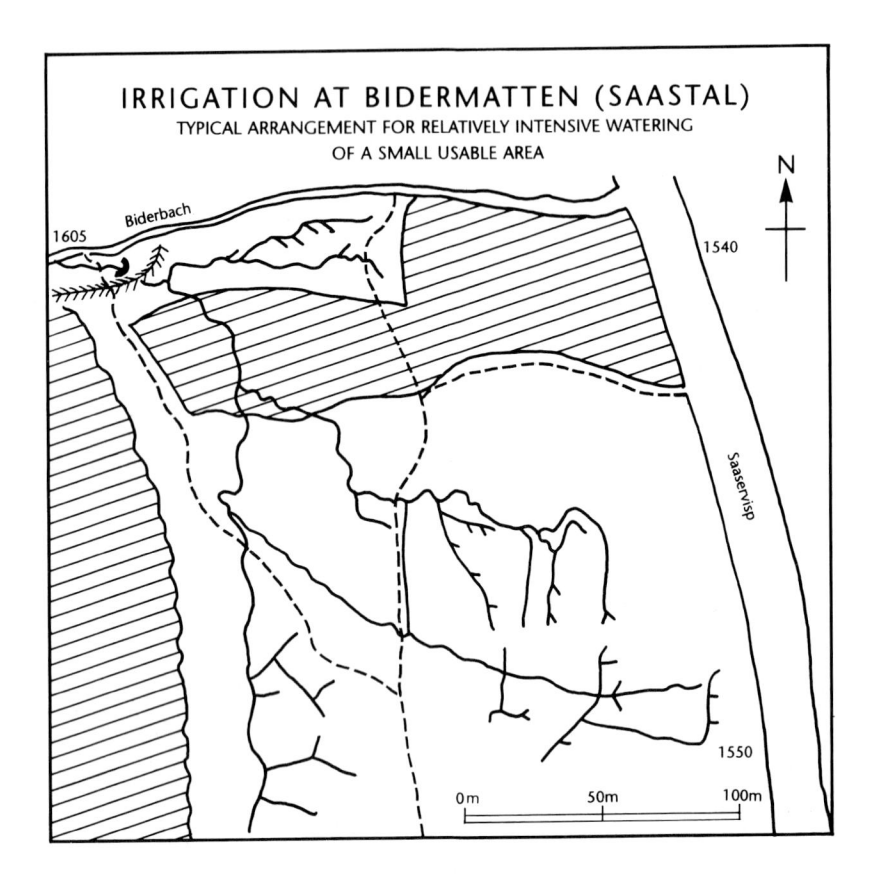

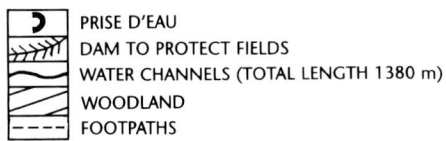

Fig. 49

Appendix

INVENTORY OF BISSES

(based on that established by the Service de l'aménagement du territoire de l'Etat du Valais.)

KEY

Act = *in active use*
O = *yes*
N = *no*

Fci = *for crops in 1992*
O = *yes*
N = *no*

Tui = *touristic use in 1992*
O = *yes*
N = *no*

Tci = *types of crops in 1992*
P = *meadows*
V = *vines*
C = *fields*
A = *fruit trees*
PC = *meadows and fields*
PV = *meadows and vines*
PA = *meadows and fruit trees*
M = *mixed (more than 2 of above)*

Type irr = *type of irrigation used*
G = *gravity*
A = *aspersion (spray)*
M = *mixed (gravity & spray)*

Type org = *type of management*
C = *consortage*
A = *administration by commune*
P = *private, other*

Cl. = *Classification of value by SAT 1993*
L = *local*
R = *regional*
C = *cantonal*

Appendix

Bisse/Suon	Source	Alt. Source	Alt. Min	Area Watered
Eggerin	Distelbach			Bellwald
Bregerin	Wyssa			Fiesch
Laxerwyssa	Fieschergletscher			Lax, Deisch
Bergerin	Glingelwasser	2450		Martisberg, Deisch
Oberriederi	Marjälesee, Aletschgl.			Ried-Mörel
Riederi	Massa			Ried-Mörel
Obere Neue	Gamsa			Visperterminen
Blyschi	Blyschbach	2100	1500	Ergisch
Untere Riederi	Welschbächli	1100		Staldenried
Lauinsuon	Faflerbach	2000		Blatten
Brächtschunsuon	Gisenlella	1950		Blatten
Sarrasin	Torrent de Pinsec	1990	1500	St-Jean, Chalais
Neuf, Benou	Torrent des Moulins	1800	1640	St-Luc
Lacher	Torrent des Moulins	1900	1660	Ayer, St-Luc
Ro, Luyston	Ertentse	1733	1444	Icogne, Lens
Audannes	Torrent des Audannes	2520	1700	Ayent
Tsa-Cretta	Torrent de la Manna	1790	1615	Mase
Emaya, Emega	Dixence	1580	1460	Hérémence
Fang	Dixence	1060	880	Hérémence, Vex
Hérémence, Gd Trait, Gd Bisse	Dixence	1540	1460	Hérémence, Vex
Chervaix, Chervé	Printse	2370	2070	Nendaz, Veysonnaz, Les Agettes
Saxon	Printse	1850	1511	Nendaz, Isérables, Saxon, Riddes
Patier	Losentse	1500	1350	Chamoson, Leytron
Waldwasser	Milibach	1350	1300	Mühlebach
Wuhr	Milibach	1200	1140	Ernen
Bellwalderin, Fürgangerin	Am Setzenhorn, Schranni	1850	1550	Bellwald
Fiescherwyssa	Weisswasser	1230	1150	Fiesch
Dorfera	Milibach	1300	1250	Ernen
Eggera	Ausgleichsbecken "Frid"			Ernen
Milleri	Milibach	1200	1150	Grengiols
Nessgerin	Kelchbach	2100	900	Naters, Birgisch
Stockeri	Kelchbach	1057	990	Naters
Bitscheri	Massa	1000	850	Bitsch
Obere Bitscheri	Kelchbach	1150	1000	Naters
Branderi	Kelchbach	950	700	Naters
Grosse	Mundbach	1250	1130	Birgisch

APPENDIX

Length	Date constr.	Act	Fci	Tui	Type irr.	Type org.	Cl	Remarks
7.8	1687	N	N	N				
4.5	XIV cent.	N	N	O				
	XV cent.	N	N	O				
	1391	N	N	N				Wooden channel
		N	N	N				Wooden channel
7	1286	N	N	O				
	bef.1682	N	N	N				Wooden channel for 120m
4		N	N	N	P	G	P	
		N	N	N				
3		N	N	N	P	G	C	
2.5		N	N	N	P	G	C	
8		N	N	N			R	
3.5		N	N	N			L	
4		N	N	O			L	
9.5	1300	N	N	O			R	Traces of supports
8.5		N	N	O			R	Traces
3		N	N	O			L	Stonework, troughs
9	XVI cent.	N	N	O			R	Channels
7.5	1824	N	N	O			R	
10	1440	N	N	O		C	R	Traces
15	1862	N	N	O			R	
32	1865-89	N	N	O			C	Traces of supports
3.5	1817	N	N	O			L	
1.1		O	O	N	P	M	A	L
3	1378	O	O	N	P	G	A	R
	1371	O	O	N	P	G	A	R
	XIII cent.	O	O	O	P	G	A	C
3.3	XIII cent.	O	O	N	P	G	A	R
2		O	O	N	V	A	C	C
2		O	O	N	P	M	C	L
8		O	O	O	P	G	A	C
3		O	O	N	P	G	C	L
5	1924	O	O	N	P	G	A	R
2		O	O	N	P	G	A	L
3		O	O	N	P	G	A	R
4		O	O	N	P	M	C	C

Bisse/Suon	Source	Alt. Source	Alt. Min	Area Watered
Oberste	Mundbach	1300	1260	Birgisch
Wyssa	Mundbach	1550	1473	Mund
Niwa	Mundbach	1470	1410	Mund
Stelwasser	Mundbach	1410	1407	Mund, Wasen
Mittelwasser	Mundbach	1370	1337	Mund, Gstein
Stigwasser	Mundbach	13.10	1260	Mund, Sitte
Dorfwasser	Mundbach	1310	1055	Mund, Gstein
Badneri	Mundbach	1210	1100	Mund, Hohfled
Haslery	Steinenbach			Ried-Brig
Niwa	Salttina			Ried-Brig
Rischeri	Steinenbach			Ried-Brig
Bergwasser	Schlessbach	2000	1920	Ried-Brig, Termen
Elsterli	Schlessbach	1700	1650	Ried-Brig, Termen
Giebjeri	Schlessbach	2000	1930	Ried-Brig, Termen
Obere Brigeri	Saltina	850	820	Brig
Untere Brigeri	Saltina	1600		Brig
Holzerly	Saltina	855	750	Glis
Untere Neue	Gamsa			Visperterminen
Visperi	Gamsa	830	690	Visp
Eyholzeri	Gamsa	830	700	Eyholz
Heidenwasserleitung	Gamsa	2500	2212	Visperterminen
Mühlebacheri	Breitenbach	850	800	Stalden
Gasperi	Baltschiederbach	1216	1131	Eggerberg
Eggeri	Baltschiederbach	1099	1026	Eggerberg
Laldneri	Baltschiederbach	900	800	Eggerberg, Lalden
Tenneri	Baltschiederbach	916	797	Eggerberg, Lalden
Äbneri	Baltschiederbach	1013	900	Eggerberg
Niwärch	Baltschiederbach	1287	1259	Ausserberg, Gründen
Mittla	Baltschiederbach	1287	1127	Ausserberg, Gründen
Undra	Baltschiederbach	1099	1046	Ausserberg, Gründen
Wiingartneri	Baltschiederbach	916	788	Baltschieder
Manera	Bietschbach	1056	945	Raron, St-German
Niwa	Bietschbach	860	640	Raron, St-German
Grossa	Jolibach	1000	900	Raron, St-German
Chumwasser	Bietschbach	1150	1000	Raron, Niedergestein
Ladu-Suon	Jolibach	1939	1400	Hohtenn
Glesch-Tatz-Suon	Jolibach	1750	1080	Tatz, Glesch
Lüungärru	Jolibach	931	915	Niedergestein, Hohten
Stägäru Suon	Jolibach	900	850	Hohtenn
Obere Wasserleitung	Lonza	700	650	Gampel
Tschingel	Jolibach	2200		Gampel

Length	Date constr.	Act	Fci	Tui	Tci	Type irr.	Type org.	Cl	Remarks
3.5		O	O	N	P	M		C	
1.5	XIV cent.	O	O	N	P	G	C	C	
1.5	1521	O	O	N	P	G	C	R	
2	1333	O	O	N	P	G	C	L	
1.5	1418	O	O	N	P	G	C	L	
	1521	O	O	N	P	G	C	L	
	1297	O	O	N	P	G	C	L	
6.4	1513	O	O	N	P	G	C	C	
	1805	O	O	N	P	G	A	L	⎫
	1805	O	O	N	P	G	A	L	⎬ Largely in tunnels
	1902	O	O	N	P	G	A	L	⎭
7.5	1938	O	O	O	P	G	A	C	
3	XVIII cent.	O	O	N	P	G	A	L	
	1650	O	O	N	P	G	A	L	
	XVI cent.	O	O	N	P	M	C	R	
2.5		O	O	N	P	M	A	R	
2.8	XVIII cent.	O	O	N	P	M	C	R	
	1500	O	O	N					⎫
6.5		O	O	O	P	A	A	R	⎬ Sections in wooden channels
2.5		O	O	N	P	A	A	L	⎭
15	1305	O	O	O	P	G	A	C	
1.5		O	O	N	P	G	A	R	
	1640	O	O	O	P	M	A	C	
3		O	O	N	P	M	A	L	
7		O	O	O	P	M	A	C	
3		O	O	N	P	M	A	L	
1		O	O	O	P	M	A	L	
8.5	1381	O	O	O	P	M	P	C	⎫ partly in tunnel
5.5		O	O	N	P	M	A	L	⎭
6.5	1377	O	O	O	P	M	A	C	
4		O	O	N	PC	M	C	R	
4	1300	O	O	O	PV	M	A	C	
2.5		O	O	N	PV	M	A	L	
1		O	O	O	P	G	A	L	
2		O	O	O	PV	G	A	R	
6		O	O	O	P	G	A	C	partly in wooden channel
6		O	O	N	P	G	A	L	
3.5		O	O	N	PV	M		C	
1.5		O	O	N	PV	G	P	R	
3	1900	O	O	N	V	A	A	L	
		O	O	O	PV	M	C	L	

121

Bisse/Suon	Source	Alt. Source	Alt. Min	Area Watered
Sebinettenrüss	Von Tschingel	1050		Bratsch
Leitemrüss, Chimattenrüss	Von Tschingel	1050		Bratsch
Engerschwasser		2100		Bratsch
Millwasser und Bijbrunnen	Bijbrunnen	1050		Bratsch
Riederi	Turtmänna	770	700	Turtmann, Ried
Tenneri	Turtmänna	900	800	Turtmann, Ried
Fätschi	Turtmänna	1000	700	Turtmann, Fätsch
Turtmänna	Turtmänna	1300	1200	Ergisch
Alte Suon	Millbach	1650	1400	Elscholl
Meigger Suon	Milibach	1820		Unterbäch
Wilde Suon	Milibach	1710	1100	Unterbäch
Finile-, Biel Suon	Unterbächnerin	1580	1200	Unterbäch
Alte Suon	Milibach	1880		Unterbäch, Bürchen, Zeneggen
Augstbordwasserleitung	Embdbach	2050		Zeneggen
Springerin	Törbelbach	1964	1800	Törbel
Juöseri	Törbelbach, Embdbach	1050	1000	Stalden
Hasleri, Staldneri	Embdbach	1600	1400	Embd, Törbel
Gsponeri	Siwibach	2600	1600	Gspon
Obere Riederi	Riedbach	1660	1300	Staldenried
Finileri	Siwibach	2600	1900	Staldenried
Staldneri	Vispa	1000	900	Stalden
Eya Wasserleita	Vispa	1200	1000	Eisten
Eggeri	Riedbach	1800	1750	Grächen
Kilcheri	Riedbach	1800		Grächen
Drieri	Riedbach	1800		Grächen
Bineri	Riedbach	1700		Grächen
Mattwasserleita	Riedbach	1600		St-Niklaus
Sparu Wasserleita	Jungbach	2200		St-Niklaus
Rittneri	Riedbach	1600	900	St-Niklaus
Riederi	Riedbach	1600		St-Niklaus
Niwa	Riedbach	1400		St-Niklaus
Wichjeri	Riedbach	1200		St-Niklaus
Hellenen Wasser	Riedbach	1600		St-Niklaus
Stock-Tennjen Hubuiti	Riedbach	1500		St-Niklaus
Feldwasser	Riedbach	1200		St-Niklaus
Lammwasser	Geistriftbach	1400		St-Niklaus, Herbriggen
Schwidernu Wasserleite	Blattbach	1350		St-Niklaus
Balmen Wasserleite	Jungbach	1200		St-Niklaus
Irmänzu Wasser	Jungbach	1200		St-Niklaus
Haltwasser	Täschbach	1500	1450	Täsch
Z'Oberst Wasser	Täschbach	1950	1700	Täsch
Vordersand Wasser	Mattervispa	1450	1440	Täsch

Length	Date constr.	Act	Fci	Tui	Tci	Type irr.	Type org.	Cl	Remarks
		O	O	O	P	A	C	L	
7.5		O	O	O	P	G	C	L	
		O	O	O	PC	M	C	L	
		O	O	O	PV	G	C	L	
2		O	O	N	P	G	P	L	
3.5	1895	O	O	N	P	M	A	R	
3.5		O	O	N	P	G	P	L	
6		O	O	O	P	G	C	C	
6		O	O	N	P	G	P	R	
		O	O	O	P	G	A	R	metal channel
		O	O	O	P	G		R	
		O	O	N	P	G	A	L	
		O	O	O	P	G	A	C	
12	XIII cent.	O	O	O	P	G	C	C	
3.5		O	O	N	P	G	C	L	
2.5	1439	O	O	N	P	G	A	L	
5.5	XIII cent.	O	O	N	P	M	A	C	
10		O	O	N	P	G	C	R	
4		O	O	N	P	G	C	R	300m. in wooden channel
6		O	O	O	P	G	C	R	
7	1923	O	O	N	P	G	A	R	pipe
1.2	1940	O	O	N	P	G	C	L	
6	1603	O	O	O	P	G	A	C	
3		O	O	O	P	G	C	R	
3		O	O	O	P	G	C	L	
3		O	O	O	P	G	C	R	
3.5		O	O	O	P	G	C	R	
2		O	O	O	P	G	C	R	
3.5		O	O	N	P	G	C	L	
2		O	O	N	P	G	C	L	
3.5	1960	O	O	N	P	M	C	L	
1.5		O	O	N	P	G	P	L	
2		O	O	N	P	G	C	L	
2.5		O	O	N	P	G	C	L	
1		O	O	N	P	G	C	L	
1.5		O	O	N	P	G	C	L	
2		O	O	O	P	G	C	L	
1		O	O	N	P	G	C	L	
1		O	O	N	P	G		L	
1.2	1992	O	O	O	P	G	A	L	
2		O	O	N	P	G	C	L	
1		O	O	N	P	G	C	L	

Bisse/Suon	Source	Alt. Source	Alt. Min	Area Watered
Hintersand Wasser	Mattervispa, Täschbach	1440	1420	Täsch
Neu-Wasser	Täschbach	1500	1430	Täsch
Blaswasser	Täschbach	2000	1500	Täsch
Giessen	Turtmänna	1000	600	Turtmann, Agarn, Leuk
Meretschi	Meretschlbach	790	650	Leuk. Agarn
Alte Leuker Suon	Turtmänna	1000	800	Leuk
Illty	Illsee	2360	750	Leuk
Raspillwasserleitung	Raspille	1100	760	Salgesch, Varen
Dala Wasserleitung	Dala	900	800	Leuk, Varen
Bru	Fonte de neige	2400		Erschmatt
Altes Wasser	Mattwaldbach	2320		Saas Balen, Eisten
Fellbachwasser	Fellbach	1940		Saas Balen
Wiessu Bäch	Wiessu Bäch	1940		Saas Balen
Almagellerbach-Furusand	Allmagellerbach	1780	1700	Saas Allmagel/Grund
Almagell-Zer Brigge	Allmagellerbach	1750	1700	Saas Allmagel
Zermeigger-Alpien	Meiggerbach	1750	1700	Saas Allmagel
Chrumbacheri	Chrumbach	1590	1480	Simplon
Halbstalleri	Lowigraben	1500	1350	Simplon
Stadleri	Senggibach	1790	1600	Simplon
Chluisbieri	Lowigrabe	1500	1350	Simplon
Engi Wasserleita	Walibach	1800	1700	Simplon
Gassensuon	Gisenlella	1940		Blatten
Oberriedsuon	Gisenlella	1940		Blatten
Mühlisuon	Lonza	1450		Blatten
Torbun Suon		1800	1700	Ferden
Alpig Suon		1800	1400	Ferden
Haselleen Suon		2100	1300	Ferden
Oberdorf Suon		1450	1400	Ferden
Vercorin, Gd Bisse	Rèche	1680	1400	Chalais
Brie	Trop-plein usine Navisence	1100	880	St-Luc, Chalais
Ricard, Chararogne	Navisence	800	600	·Chalais
Rèche	Rèche	620	520	Chalais
Orméan	Rèche	620	580	Réchy
Fleur	Rèche	540	520	Réchy
Etreys		510	504	Gröne
Neuf	Rèche	1180	1080	Gröne
Morété, Morestel	Rèche	1226	980	Gröne
Gröne	Rèche	1226	600	Gröne
Marais	Raspille	620	540	Bernunes d'en bas
Planige	Raspille	1100	920	Mollens, Venthône, Veyras
Tsittoret, Zittoret	Tièche	1960	1720	Mollens, Randogne, Venthône, Veyras

Length	Date constr.	Act	Fci	Tui	Tci	Type irr.	Type org.	Cl	Remarks
1.5		O	O	N	P	G	C	L	
1.2		O	O	N	P	G	C	L	
2.5		O	O	N	P	G	C	R	
3		O	O	N	P	G	C	L	
2		O	O	N	P	G	C	L	
	1900	O	O	N	P	G	A	R	
2		O	O	N	P	G	C	L	
4.5		O	O	N	PV	M	A	C	
1.5		O	O	N	P	M	A	R	
		O	O	N	P	G		R	
		O	O	O	P	A	C	L	
	1991	O	O	O	P	M	C	L	
		O	O	N	P	G	C	L	
1		O	O	O	P	G	A	L	
2		O	O	O	P	G	A	R	
2		O	O	O	P	G	A	L	
3.7		O	O	N	P	A	A	C	
1.3		O	O	N	P	A	A	L	
1.6		O	O	N	P	G	A	L	
1		O	O	N	P	A	A	L	
1.5		O	O	N	P	M	C	L	
2.8		O	O	O	P	G	C	L	
2		O	O	N	P	G	C	L	
1		O	O	N	P	G	C	L	
1	XIX cent.	O	O	N	P	G	C	L	
3	XIX cent.	O	O	N	P	G	C	L	
4	XIX cent.	O	O	N	P	G	C	L	
1	XIX cent.	O	O	N	P	G	C	L	
6	1358	O	O	O	PV	M	C	C	
3.5	1923	O	O	O	PC	M	C	R	
7	1484	O	O	O	M	M	C	C	
2		O	O	N	A	A	C	L	
1.5	1903	O	O	N	V	A	C	L	
1	1937	O	O	N	PV	A	P	L	
1	XIV cent.	O	O	N	A	A	A	R	
4	XVI cent.	O	O	O	P	G	A	R	
2.5	XIV cent.	O	O	N	P	G	A	R	
5	XIV cent.	O	O	O	PA	M	C	R	
2.2		O	O	O	V	M	C	L	
7.5	1400	O	O	O	PV	M	A	C	
8	1400	O	O	O	PV	A	A	C	

Bisse/Suon	Source	Alt. Source	Alt. Min	Area Watered
Gd Bisse de Lens, Riouta	Lienne	1170	960	Icogne, Lens, Montana, Chermignon
Sillonin, St-Léonin	Lienne	950	550	Icogne, St-Léonard, Lens
Clavau, Clavoz	Lienne	680	520	Ayent, Grimisuat, Sion
Ayent, Bisse Neuf, Gd Bisse	Lienne	1520	940	Ayent, Arbaz, Grimisuat
Bitailla, Taillaz	Sionne	1510	1220	Arbaz, Ayent
Sion	Lienne	1820	1120	Sion
Grimisuat	Sionne	1214	940	Arbaz, Grimisuat, Sion
Déjour et Bourzi	Sionne	1294	940	Savièse
Lentine	Sionne	760	643	Sion, Savièse
Tsampé, Fontanay	Sionne et sources et tunnel	1430	1140	Savièse
Torrent Neuf, Ste-Marguerite	Nettage, Morge	1400	1020	Savièse
Montorge	Lac de Montorge	643	540	Sion
Tsandra	Morge	1430	930	Conthey
Champys	Lizerne	780	550	Ardon
Euseigne	Dixence	1200	1160	Euseigne
La Muraz	Chaulué	2280	1460	Hérémence, Vex
Vex	Printse	1520	1300	Nendaz, Veysonnaz, Agettes, Vex
Salins, Gd Bisse de Salins	Printse	1210	600	Nendaz, Salins
Baar	Printse	960	720	Nendaz, Salins, Sion
Bisse d'En Bas, de Dessous	Printse	1390	950	Nendaz
Bisse du Milieu	Printse	1440	1350	Nendaz
Bisse Vieux, d'En Haut	Printse	1560	1400	Nendaz
Levron	Torrent de Versegères	2300	1500	Vollèges
Meunières de Bramois	Borgne	496	492	Bramois
Meunières de Champsec	Borgne	510	500	Sion
Siphon de la Muraz	Lac de Montorge	670	600	Sion
Charrat, Canal du Guercet	Drance	500	450	Charrat
Meunières de Martigny	Drance, Trient	500	460	Martigny
Trient, Martigny-Combe	Trient	1583	1520	Trient, Martigny
Verney	Torr. d'Aron	2260	1470	Liddes (Chardonne)

Length	Date constr.	Act	Fci	Tui	Tci	Type irr.	Type org.	Cl	Remarks
13.8	1448	O	O	O	M	M		C	
7.5	1367	O	O	O	PV	M	C	C	0.6 ct/m²
7.7	1453	O	O	O	V	A	A	C	masonry walls and supports
15	1442	O	O	O	M	M	C	C	restored section of wooden channel
4	bef. 1307	O	O	O	M	A	A	C	
13.5	1903	O	O	O	V	A	A	C	pipe in 1972; drinking water Anzère
3	XVIII cent.	O	O	O	PV	A	A	R	
4.5	bef. 1667	O	O	N	P	G	C	R	
4.5		O	O	O	V	G	A	R	4 ct/m²
5.5		O	O	O	P	G	C	R	
11	bef. 1430	O	O	O	PV	M	C	C	braces, boutsets
2.5	1885	O	O	O	V	A	C	R	4 ct/m²
11.7	1400	O	O	O	PV	M	A	C	
3	1860	O	O	N	V	A	A	L	
3.6		O	O	O	P	G	C	R	
8		O	N	O				R	
12	1453	O	O	O	PC	A	C	C	
12	1435	O	O	O	M	M	C	C	
6	1456	O	O	O	PA	M	C	C	
6		O	O	O	M	M	C	C	
5	1700	O	O	O	A	M	C	C	
7		O	O	O	PC	M	C	C	
13	1465	O	N	O				C	Replaced by Louvïe canal
4		O	O	N	M	M	A	R	
15	very old	O	O	N	PA	G	A	C	5Fr./1000 m²
1	1934	O	O	O	V	A	C	R	
6.5	1847	O	O	N	A	M	A	R	
4		O	O	N	M	M	A	L	about 10 channels; no charge for use
4		O	O	O	A	G		R	
3		O	O	N	A	A		L	

Glossary

Ablation	Erosion or removal; chiefly used to describe the reduction of glacial ice to water vapour.
Ardjou, (g)erdiou, Arzieu, Azieu	Officially appointed waterer
Aspersion	Watering by overhead sprinkler
Avoyour	President of the "Commission du bisse" who supervises the migniours, q.v. (Lens)
Banquette	Walk-way (along a bisse)
Barrate	Water-hammer (audible device indicating continuation of flow)
Bâton à marques	Wooden staff on which water rights are recorded
Bazot	Section of wooden channel for a bisse, usually a hollowed-out tree trunk, but can be built of planks or sheet metal (Sion)
Bedarra	Small channel
Boutzet	A beam, preferably made of larch, 20 cms square in section, supporting the wooden channel of a bisse (e.g. across a rock face); usually marked with the year installed and the name or badge of the man who installed it or paid for the installation
Boutzesse	A square hole 15-20 cm. deep in the rock into which a boutzet is inserted and made fast with larch wedges
Brochet, brotzet	See bazot (Bas Valais)
Bulletin	(i) square hole in a répartiteur (q.v.) (ii) unit of flow (iii) water-right
Butset	See boutzet
Carrelet	Square-section beam
Celluse	See écluse
Chänil, chenal	See bazot
Cheville	Wedge (used in bisse construction)
Choppa	Shutting of sluices
Chrapfen	See Krapfen (ii)
Clayonnage	Prise d'eau made out of tree-trunks

Commune	Area of local government in Valais which often includes a number of separate villages or settlements
Componction	Right to extend one watering period into the next if a bisse is not flowing at full capacity (Lens)
Computiste	"Numbers man" who does any arithmetic required in connection with a bisse (Savièse)
Console	Bracket
Consort	Member of a consortage
Consortage	Association of individuels who own, maintain and operate a bisse (in contrast to when this is done by a commune)
Co-outacoué	Vertical beam connecting two horizontal ones in the woodwork supporting a bisse crossing a rock face (Savièse)
Cria	General work on a bisse (Lens)
Criées	Public announcements
Décharge du bisse	Emptying a bisse by blocking the entrance to it, usually with a sluice, at its prise d'eau and thus directing all the water down the torrent
Déchargeoir	Large sluice and channel through which a bisse can be directed if it is necessary, e.g. on account of a break, to prevent water continuing downstream
Délabre	Broad-bladed pick-axe for opening and closing holes in the bank of a bisse or smaller distribution channel to irrigate land
Dépantse	Overflow
Dépotoir, désableur, désabloir	Sand trap
Destournyour, détournoir	Sluice for diverting the flow of a bisse on to a piece of land (Lens)
Déversoir	Area, e.g. a pond, into which a bisse can be diverted, e.g. on Sundays, when watering was forbidden
Diviseur	Distribution point, where the flow of a bisse is divided into various channels
Droit	Amount of water which an entitled party can draw, e.g. 50 litres a second
Écluse. enclorse	Sluice: a device for controlling or stopping the flow of water either along or out of a bisse. In its most primitive form, a flat stone; then a flat stone with a square hole in it related in size to the capacity of the bisse; then a wooden or metal, vertically-lifting shutter in a concrete or metal mounting and sometimes fitted with a padlock.

Encorbellement	Bracket, cantilevered support for a wooden channel
Erdjiou, erzyu	Waterer (Savièse), see also Ardjou
Erzère	Small channel
Erwin, ewin	Man appointed to allocate water
Étanche	Metal (or, rarely, wooden) plate pushed diagonally into a bisse or smaller channel to deflect the water through a hole in the bank on to land to be irrigated
Étau	Vertical beam, see co-outacoué
Fauchée	A quantity of grass cut with a scythe
Firn	Granular ice which gradually develops from the original snow in an area where a glacier starts. Also known as névé
Ganglatte	Wooden walk-way over or beside a bisse
Garde, garden, gardien,	Guard whose function is to patrol a bisse and carry out maintenance and minor repairs
Géto	See étau and co-outacoué
Générale	General repairs carried out every spring before the levée (q.v.) of a bisse (Le Levron)
Gouille	Pond
Hüter, Wasserhüter	See garde
Kalafra, Kalatra	See sluice
Känel, Kennel	See bazot
Känelkrapfen	Naturally-grown hook-shaped brackets for carrying bazots (q.v.) across rock-faces
Kontrollhammer	See barrate
Kapiou	i) sluice ii) small channel
Kehr, Wasserkehr	Time necessary for watering all the land which a bisse serves and the order in which those entitled take their shares
Krapfen	i) wooden pole with an iron hook on one end for controlling a bazot (q.v.) when it is being lowered into position (Ausserberg) ii) naturally hook-shaped piece of wood, fixed with a wedge to a boutzet (q.v.), to support a bazot (q.v.)
Kusi	i) See étanche ii) small channel
Légotuire, ligotuire	Abandoned or excess water
Levée du bisse	Starting the bisse flowing by letting the water into it at the prise d'eau

Leviour	Prise d'eau
Levoir	Small channel
Mayen	Small barn for the storage of hay or grain and, by extension, a summer pasture with a number of these on it.
Mechuire, messuire	Poses (q.v.) sold to individuals who have not sufficient water-rights to irrigate all their property. Usually three poses each week (Lens)
Merkhammer	See barrate
Métral	Foreman
Meunière	Mill-leat
Mièze	Odd-job man
Migniour	Official who, under the orders of the avoyour (q.v.), selects men for the annual and special work on a bisse, supervises it, keeps the accounts for it and sells the mechuires (q.v.) (Lens)
Mise en charge	See levée du bisse
Moneresse	Small channel
Monaz, munez	Covered bisse in a steeply-descending pipe
Moulinet	See barrate
Oétan	Pond
Parchet	Place, area
Partichoux, partitiou	See diviseur
Pas de la Matta	See page 34
Passage obligatoire	Unavoidable difficult or dangerous passage in the course of a bisse
Pioche	Broad-bladed pick-axe (see délabre)
Plaque d'arrosage	See étanche
Ploton	Cover, made of logs, for a bisse in an area subject to falls or slides of débris (Le Levron)
Poin plan	Two beams fixed at right-angles to each other, bracket
Pose	Water of a bissse for a certain, usually short, time (Savièse, Lens)
Prise d'eau	Point of access to a torrent where a bisse starts
Procureur	Assistant supervisor of the control and maintenance of a bisse (the functions vary; in some cases he is the equivalent of avoyour (q.v.), in others his duties are much less onerous)

Quartier	Division of a commune or of an area served by a bisse
Raie (reà, ri, ru)	Local words for bisse in Drance valley
Raie coursière	Main channel of a local irrigation arrangement, drawing water from a torrent or bisse and taking it some distance before distributing it through small channels
Ravin	Destructive uncontrolled flow of water, e.g. from a break or overflow of a bisse
Recteur	See procureur (Le Levron)
Répartiteur	See diviseur
Rigole	Small channel
Romoiée	Supplementary poses (q.v.) at the start of the tour (q.v.) of each quartier (q.v.). It is not of the same value as the other poses because time is lost switching the water from one sluice to another
Rottenschlag	A channel carrying surplus water from the lower end of an irrigated area down to the Rhone (Rotten in Swiss-German)
Roue à paletttes	See barrate
Ru	Irrigation channel (Val d'Aosta)
Ruissellage, ruisellement	Irrigation by allowing water from a bisse or smaller channel to run downhill on the surface of the land
Sabloir	See désableur
Sander	i) see désableur ii) the man who empties the sand-traps and, by extension, patrols and maintains a suon (q.v.)
Sandfang	See désableur
Saut de la Matta	See page 34
Seiteur	An area of grass that a man can cut with a scythe in three hours
Schöpfi	See prise d'eau
Soppaz	See choppa
Speicher	See mayen (without the extension)
Stadel	See mayen (without the extension, but specifically one built on round flat stones to stop rats getting in)
Strahlen	To dig crystals out of rock
Suon	Bisse (in some parts of Haut Valais)
Tablat	Terrace in vineyard
Tasela, tasera	Wooden measure

Tassoz	Section of a bisse in a quartier (q.v.) (Lens)
Tessel, Tessle, Tässlu	Wooden token marked with the house badge and entitlement to water-rights of the owner
Tête de bisse	See prise d'eau
Tindroü	Support of a walk-way (Le Levron)
Toggenloch	See boutzesse
Toise	A measure of length
Tora	Main channel
Torgnîo	Cross-roads
Torieu	See étanche (Martigny)
Torgnoü	See étanche (Le Levron)
Tornieu, torniou	See étanche (Bourg)
Tour	See Kehr
Tourniquet avertisseur	See barrate
Tournoir, tornyeü	Small bisse
Trait	Bisse (Hérémence)
Trajeu, trajoü,	Bisse (Le Levron)
Tranchant	See étanche
Trasoir, traysiour	Main channel
Trazoé	Small channel
Tregeux	Gardien of a pond (Zeur de Mirieuse)
Tretschbord	See banquette
Tsaseila	Small channel
Tséna, tsénei	Channel made of planks
Viou	i) small channel ii) lateral bisse
Vouasseur	"Paddler", in the Bisse de Savièse
Wässerbeil	See pioche
Wasserbüchlein	Record of water-rights
Wasserfassung	See prise d'eau
Wasserfuhr	See Suon
Wasserhammer	See barrate
Wasserhüter	See garde
Wässerhaue	See pioche

Wasserkehr	See Kehr
Wasserknebel	See bâton à marques
Wasserleite	See suon
Wässerplatte	See étanche
Wasserrad	See barrate
Wasserscheit	See bâton à marques
Wasserschlegel	See barrate
Wassertessel, Wassertessle,	See Tessel
Wasservogt	See garde

ACKNOWLEDGEMENTS

I should like to thank all those who have helped me, in their various ways, to produce this book and, in particular, the following: my wife Françoise, for being consistently wifely, for walking many of the bisses with me, for taking many photographs and for putting up for the last year with even greater than usual piles of paper, books and maps all over the house; our three children, for their interest, enthusiasm and encouragement to get on with it – Nick has taken photographs and underwritten the cost of publication, Jeremy and his wife Mónica have designed and styled the book and Jocelyn, after floating various "boats" down bisses years ago, has made appropriate daughterly noises on the subject ever since; George Behrend, old friend and publisher (Jersey Artists), for all sorts of advice to an elderly tyro; Caroline Griffith for greatly improving what was initially a rather poor chapter on geography, for some very constructive editorial comment and for converting, with the aid of Fred Sanyo, her totally reliable but unimaginative electronic assistant, my manuscript into a form in which it could be used by my helpful printers; Jean Travelletti, Secretary of the Commune of Ayent and Firmin Morard, President of the Grand Bisse d'Ayent for a great deal of help with details and documentation about the bisses in the Ayent area; Roland Varone, Secretary of the Commune de Savièse, Christian Dessimoz and his colleague of the Services Techniques of the Commune of Conthey and Jean-Luc Rey of the Services Industriels de Sion for similar assistance for their respective areas; Emmanuel Reynard of the Department of Geography at the University of Lausanne for invaluable help with names and statistics of bisses and climate; Darren Crook of the Department of Geography at the University of Huddersfield for some documentation which he had found and I hadn't; M. M. Chrimes of the Institute of Civil Engineers in London; Dr. Med. Hans Hug of Basel, for pointing me in the right direction to find answers to some of my enquiries; Lily Fischer of Luzern, for ensuring that my efforts have received more publicity than would otherwise have been the case; Heidi Reisz of the London Branch of the Swiss National Tourist Office and Dr. G. Kneubühl of the Schweizerisches Alpines-Museum in Bern for information on glaciers; two cousins-by-marriage, Michel Burnand, for finding some documentation and attending to sales in Switzerland, and Jean-Marc Burnand, for finding some historic photographs; Jean-Luc Rouiller and other members of the very cooperative staff of the Bibliothèque Cantonale du Valais at Sion and Judith Kuhn of the Schweizerische Landesbibliothek at Bern; Jayne Dunlop, Librarian of the Royal Geographical Society; Keith Davis of the Royal Academy of Engineering for solving a problem with the word 'encorbellement'. Rose-Claire Schüle for advice on the relative absence of the bisse in Valaisan mythology; Melchior Kalbermatten of the Union Valaisanne de Tourism; Nadia Revaz of the Department de l'Instruction Publique du Canton du Valais (Activités Culturelles); Carmen Trachsel of Anzère for calling my attention to the then forthcoming First International Colloquium on Bisses in Sion in September 1994; Jean-Henry Papilloud of the Société d'Histoire du Valais Romand who organised that Colloquium, invited me to attend it and agreed to the distribution at it of a pamphlet about this book; all those who have responded to that pamphlet by indicating an interest; last but by no means least, all of you who have bought it.

Bibliography

Annaheim, Hans:	Die Wasserfuhren im Wallis, article in Leben in Umwelt 1949, pp 25-33
Bérard, Clément:	Bataille pour l'Eau, Les Editions Monographie, Sierre, 1976 (2nd edition)
Blotnitzki, Leopold:	Über die Bewässerungskanäle in den Walliser Alpen, Rieder in Simmen, 1871
Buercher-Cathrein, Catherine:	Der Letzte Sander von Oberried (novel), Brig, 1977 (reprint)
Bumann, Peter:	Les Bisses, article in Tiefbau, no. AC 13/81, Zürich
Chavan, Paul:	Contribution à l'Étude de l'Irrigation dans le Canton du Valais, Bern, 1915
Conseil Municipal de Sion:	Bisse de la Lienne, Cahier des Charges, 1901
Courthier, Louis:	Les Bisses du Valais, article in Echo des Alpes, 1920, no. 7-8
Crettaz, P. Sulpice:	La Contrée d'Ayent, Sion, 1933
Eichenberger, Ewald:	Beitrag zur Terminologie der Walliser Bisses, Aarau, 1940
Framji, K. K. and others:	Irrigation and Drainage in the World, International Commission on Irrigation and Drainage, New Delhi, 1983
Franzoni, Albert:	L'Aquaduc ou Bisse de Savièse, Geneva, 1894/1982
Gilliard, François:	Au fil de l'eau, Inspection des bisses de Clavoz, Lentine & Montorge, Sion, 1981
Grove A.T. & J.M.:	Traditional montane Irrigation Systems in Modern Europe: an Example from Valais, Switzerland, Agriculture, Ecosystems and Environment, Amsterdam, 1990
Hopfner, Georges:	Notice sur les Bisses du Valais, Ve. Congrès international d'agriculture, 1898
Jeune Chambre Economique:	Le Val de Réchy et ses Bisses, Sierre, 1982
Kleindienst & Schmid:	Règlement pour la police des Bisses, Sion, 1913
Lehmann, Louis:	L'Irrigation dans le Valais, Revue de Géographie, Vol VI, Paris, 1911
Macherel, Claude:	L'eau du glacier, Lötschental, Études Rurales, nos. 93-94, 1984
Mariétan, Ignace:	Le Bisse de Savièse, St. Maurice, 1934

Mariétan, Ignace:	La Lutte pour l'eau et la Lutte contre l'eau en Valais, Actes de la Société Helvétique des Sciences Naturelles, session 22, 1942
Mariétan, Ignace:	Heilige Wasser, Schweizer Heimatbücher no. 21/22, Bern, 1948
Paris, Charles & others:	Le Bisse du Torrent Neuf à Savièse, Lausanne, 1943
Paris, Charles & Seylaz, L:	Le Bisse de Savièse – I Torin Nou, Ecublens, 1988
Quiglia, Canon Lucien:	L'ancien Lens, Le Régime et ses eaux, Lens, 1984
Rauchenstein, Fritz:	Die Bewässerungskanäle im Kanton Wallis, article in Zeitschrift für schweizerische Statistik, 1907.
Rivaz, André de:	Bisse-Syphon de Montorge à Sion, report of 50 years of, Sion, 1945
Robert, Etienne:	Les Bisses de Saxon et du Levron, Bulletin de la Société Neuchâteloise Géographique no. 34, pp 16-26, 1925
Schnyder, Theo:	Das Wallis und seine Bewässerungsanlagen, Schweizerische Landwirtschaftliche Monatshefte no. 10-12, pp. 214-218 etc., 1924
Schmid, Felix:	Ausserberg und sein Wasser, in Winter 1960-61, 48 pp, Second edition, Visp, 1981
SAC Ortsgrupe Ausserberg:	600 Jahre Wasserleitung Niwärch, 76 pp, Brig 1981
Stelling-Michaud, Sven:	Vercorin, une commune Valaisanne au Moyen-âge, Les Bisses de Vercorin, Chalais & Réchy, Valesia, Sion, 1956
Stevenson, Douglas:	The Bisses of Valais, Canadian Geographical Journal no. 51, pp. 206-224, 1955
Valentin, André:	Conthey – mon pays, La Commune et Bourgeoisie de Conthey, Sion, 1979
Vautrier, Auguste:	Au pays des Bisses, pp. 151, Lausanne, 1928/1942
Vittoz, E & Schnegg:	Les Bisses, Nouvelles Éditions Illustrées, 1928.
Zenzuenen, A:	Von der Not und dem Segen des Wassers in Lax und Martisberg, Walliser Jahrbuch, 1961
Zollinger, Alfred, & Imboden, Adrian:	Wasserfuhren im Wallis, Beckerhof, Schweizerischer Lehrerverein.

Index

Entries in *italic* refer to black & white illustrations, maps and charts.
Entries in ***italic bold*** refer to colour plates.

Aar 16
Ablation 22
Aletschgletscher 22
Alluvial 19,33,44,45
Alte Suon (Turtmanntal) 59
Amédée, Count, of Savoy 25
Anniviers 78,98
Aosta 21,28
Anzère 22,*90*,98,103
Arbaz 27,30,56,58,71,*94*
Armeillon, Rocher d' 45
Arzieu, Azieu 47,65
Aspersion 64,77,97
Association 59
Attlas, Les 100
Audannes (Ohannes) Bisse des 30,45,82
Audannes, Lac des 45
Audey 93
Augstbordwasserleitung, Augstborderin 21,*27*, 31,*90*,108,*113*
Ausserberg 11,13,21,27, 28,37,*44*,55,59,60,66, 78,83,86,107,108,*112*
Aven 28,95,96,102
Avoyour 61,62,63
Ayent 8,30,31,46,52,54, 56,58,60,70,71,92,93, 98
Ayent, Bisse d' 13,21,23, 28,34,*45*,*48*,59,78,*84*, 85,*91*,*92*,93,105,***1***,***2***,***3***
Azerin, Bisse d' 82

Bacon, Pra 102
Badneri (suon) 21
Bagnes, Vallée de 21,22, 72,91
Baltschiedertal 27,44,54, 86,88,89,107,*112*
Banquette 37,*91*,*92*
Bärgerwissa 27,39,65,70,
Barma de la Dzour 35
Barrate 55
Bâton à marques *74*,*75*

Bazot 39
Bec-des-Roxes 92
Belalp - Nessel 21,108
Bellwald 27
Béra 57
Bérard, Clément 31,72
Bern-Lötschberg-Simplon railway 86
Bernese Alps 15,16
Bidermatten *116*
Binna 22
Bietsch, Bietschtal 27,78
Birgisch 78,108
Bishop of Sion 25,27,69
Bisse, definition of 7
Bisse sec 30
Bitailla, Bisse Taillaz 21, 27,59,71,79,103
Bitscherin *43*,54
Blatten *37*,54,108
Blignoud 58,79
Blotnitzki, Leopold 81
Blumlisalp(Thun) Section of S.A.C. 13
Bon d'irrigation 66
Boots, rubber 98
Borgne *115*
Botyre 58,79
Bourgeois 60,61,64
Bourzi, Bisse de 52
Boutzesse 39
Boutzet 39,47,51
Branlires, Paroi des 12,35, *35*,*36*,49,*50*
Brig 19,43,108,109
Brochet 39
Brotzet 39
Bulletin 55,*57*,65,68
Bürcher-Cathrein, Catherine 89
Butset 39

Cantilevered channel 35, *36*,102
Caroline, Rénonciation à la 25,26,69
Cattle's Stomachs, cold water bad for 22

Châble, Le 100
Chalais 27,29,72,78,*114*
Chamoson 82
Champlan *29*,106
Chandolin 52
Chänel, chänil 39
Chararogne 29
Chardonnay 28,72
Charlemagne 25
Châtelard, Mont du 90
Chaux, Mont-la- 28
Chelchbach 108
Chelin *90*,*8*
Chemical content of water 22,33,76
Chemin 92
Chermignon 60,75,107
Chervais, Chervé, Bisse de 21,101
Chrapfen 39
Clavoz 21,*29*,*29*,*30*,39, 85,106
Clerics 69
Cleuson, Lac de 101
Coinage 67
Collon, Col de 21
Colloquium,International on Bisses 95
Combaz, La 30,71
Commune 8,11,26,37,59, 64,67,73,93,97,98
Componction 64
Computiste 65
Cone, sedimentary or alluvial 19
Confederation 26,31
Consortage 37,59,85,93, 98
Conthey 79,95 et seq
Conthey, Mayens de 102
Co-outacoué 39
Cordel, Curé 23,69,79
Costs 31
Cotterg 72
Crans Montana 106
Crettaz, Rev. P 70
Cries 93
Croix Torrent 52,*84*,85,105

141

Croué Torin 28
Crucifix 85,*86*
Crystalline rock, schist 19,22,33

Dailley, Le *90*
Daillon 28,95
Dajon 49,51
Dark Ages 26
Dejour, Dijore, Bisse de 52,102
Décharge du bisse 97
Deischbach 70
Déchargeoir 67
Délabre 56,79
Déversoir 78
Directeur 65
Distributeur, diviseur *34*, 55,*57*,*58*
Dixence, La Grande 95
Dizains, State of the VII 25
Dorbon *34*,56,58,71
Drahin Torrent 52
Drance 25,79,93
Dreifurrenschiene 76
Droit d'eau – see water rights
Dzou, Mayens de la 102

Eaux Froides, Sixt des 45
Ebibergeri 35,82
Ecclesiastical authority 72
Ecône, Torrent d' 72
Eggen 107
Eggerberg 66,107
Eghvogerts 68
Emd 27,*113*
Emdbach 21
Encorbellement (canti-levered channel) 131, 138
Ehéley 105
Erde 28,95,96
Ergisch - Turtmann Wasserleitung 107
Ernen 27
Ertense torrent 28
Erwin 75
Etanche 56,*58*
Etau 39
Evaporation 30
Evolène 8,21

Falaj 15
Fardel, Canon 69,86
Federal Government 31
Fées, Bisse des 12
Fenêtre, Col de 21
Fey 20

Fiesch 27,66,109
Fieschergletscher 27,70
Fieschertal 70
Fieschertalsuon 109
Fille, la 30,92
Fillette, la 30,92
Fire-fighting 61,67
Firn 22
Fleives (Lombardon) 58, 79
Floods 19,76
Foehn 19,
Folk-lore 73
Fontana Dzemma 28
Forclaz, Col de la 99
Forniri, Torrent de 48,*48*
Fürgangen 27
Furka 16,52

Gamsa 21
Garde, garden, gardien, Guard 39,46,48,63
Gasperi, Gasparisuon *89*, 107
Gebidem 88
Geneva, Lake of 15
Géto 39
Giétro glacier 23
Glacier milk 22,23,33
Glaciers, movement of 22,23
  water temperature, bad for cattle 22
Gletschermilch 22,33
Glingulwasser 109
Go, Le 103
Gouille 65
Gouilles Les, (de Savièse) 65,78
Grächen 20,55
Grand Trait d'Hérémance 71
Grand-Dzou 102
Granite 19
Gredetschtal *45*,*48*,54
Grimenz 78
Grimisuat 59,60,71,78, 105
Grône 27,72
Grosser Aletschgletscher 22
Gründen 60
Gspon 76
Guntern, Joseph 12

Heidenwasserleitung 21
Hérémence 54
Hérémence, Bisse d' 21, 26,71
Hérens Col d',Val d' 21,27

Hildebrand II Jost, Bishop 26
Hugues, Bishop 25
Huiton ponds 28
Huns 26
Hüter 39

Ice Age, Litle 22
Icogne 60,98,106
Illsee 79
Interruption of flow 56
Insolation 16,*18*
Isérables 29,100

Jean, Curé Pierre 70
Jube 15

Kalpetran 108
Känel, Kennel 39
Känelkrapfen 39
Kehr 78
Krapfen 39
Kühboden 109

Lalden 78
Lax 21,39,65,70
Laxerwyssa 65
Lehmann, Dr. Louis 66,78
Leiggern, Leukron, Arnold von 28,37
Lens 12,27,28,70,78,79
Lens, Grand Bisse de (or Bisse de la Rioutaz) 21, 60 et seq,70,90,98,106,*8*
Lens, Prior Jean de 26,69
Lentine, Bisse de 21,27, 30,39,*46*,105,*4*,*5*
Leuk 30
Leukerin 30
Levada 15
Levée (du bisse) 49
Levels, maintenance of 34,37,39
Leviour 33
Levron, Le 21,23,28,34, 52,59,72,75,91,92,100
Lienne, La 27,*49*,60,70,98
Lienne, Electricité de la 92,93,95
Limestone 22,33
Lin, Col du, Pas du 92
Little Ice Age 22
Livret d'Eau 65
Lombardon 58,79
Losentse 82
Lötschental 22,75
Lourantse 104
Louvie 92
Lowigrabe 107
Loye 27,72

Luc 58

Manera 78, *112*
Maps *4,111-116*
Maret, Albert 91
Mariétan, Ignace 55,*74*
Märjelesee, Märjelewang 109
Martigny 22,25,79
Martisberg 21,39,70
Massa *43,78*
Matta, Pas de la,Saut de la 34
Mattavisp 55
Mattenbach 70
Mauvoisin 25,91,92,95
Mayens-de-Conthey 102
Mayens-de-la-Dzou 103
Mayens-de-My 102
Mayens-de-Rèchy 54,*114*
Mayens-de-Riddes 29
Mayens-de-Sion 21,*115*
Mayens-de-Vercorin *114*
Mayentzet 97
Merdenson 93
Merkhammer 55,*56*
Meschuire, messuire 62
Métral 47,65
Meunière 79
Michel, Moritz 27
Miège 71,107
Mignour 61,62,63
Milibach 27,83,*113*
Mittelbach 70
Mittla suon (Ausserberg) 78,88
Molignon *30*
Montagnier 72
Montana 20,60
Mont Fort Hut 100
Mont du Châtelard 90
Mont-la-Chaux 28
Monteiller 102
Monte Rosa 15
Montorge, Mont d'Orge, Lac de, Syphon de 29
Morard, Firmin 8
Morge 21,27,31,*35*,52,96,102
Morge, Pont de la 105
Moulinet 55,*56*
Mouzerin (Daillon) 96
Mund 21,43,*44,48*,54,75,78,108
My 28

Naters 78
Navisence 72,78
Nendaz, Basse 101
Nendaz, Haute 29,100
Nendaz, Super 101
Nendaz, Bisses de 21,82,100,101
Nessel 108
Nessjerisuon 108
Netage 27,28,49
Niederwald 66
Nidla 82
Niva (Visperterminen) 89
Niwa (Emdbach-Zeneggen) 21
Niwärch 13,21,27,*38,40,41,42*,55,77,78,82,86,87,88,*88*,98,107,*6,7*

Ober-Ems 107
Oberried 55
Oberriederin (suon) 89
Oberstwasser (Gredetschtal) *48*
Octodurum (Martigny) 25
Ohannes, (Audannes), Bisse des 30,45,82
Orméan, Bisse d' 30

Païens 21,26
Pas du Lin 92
Pas de la Matta 34
Pas de l'Ours 106
Passage obligatoire 34
Passerelle des Amours *49*
Patois 8,28
Permeability 19
Pétouse, La 52,102
Pierre Avoua 91,92
Pioche 56,*58*,79
Pipes 44,91
Planches, Col des 100
Plan des Roses 45
Planchouet 100,101
Plans Mayens 106
Planta, Battle of La 26
Plaque d'arrosage 79
Ploton 47
Poin plan 39
Ponds 28,65,78
Pont-de-la-Morge 106
Pose 64,65
Pra Bacon 102
Prabé tunnel 28,35,47,*50,52,53*,89
Pra Combère 34,105
Precipitation data 16,18,*20*
Premploz 28,95,96,97
President 65
Printse 29,82,115
Prise d'eau 7,33,*34,4*
Pro Catroué *94*,104,105
Procureur 28,47,65

Proprija, La 107

Q'Anat 15
Quaglia, Canon Lucien 70
Quarter,Quartier 26,60 et seq

Raaft 82
Rain shadow effect 16,*17*
Rarogne, Raron 76
Raspille 13,27,71
Raspillewasserleitung 107
Rauchenstein, Fritz 26,81
Ravouéné 104
Rawyl 45,92
Rèche 30
Réchy 27,29,30,54,72,*114*
Rénonciation à la Caroline 25,26,69
Répartiteur, répartitour 55
Reynard, Emmanuel 13, 81,83
Rhone 15,16,19,21,25,27,79
Rhone Glacier 15,23
Ric(c)ard, Bisse de 29
Riddes 72,100
Riddes, Mayens de 29
Ried,Riedbach 55
Ried-Brig 30
Riederhorn tunnel 89
Riederin *43*,82
Ried-Mörel 78,89
Riouta(z),Grand Bisse de la 21,28,54,59,60,69,70
Rischeri 30
Ro, Roh, Bisse de 22,27,28,*49*,106
Rocher d'Armeillon 45
Rocher, Rosset, Marie 12
Rock formations 19
Rodolphe III of Burgundy 25
Romans 26
Roses, Plan des 45
Rousses, Les 22,*91*
Ruissellage, ruisellement 64,77,97
Ruodpert 25

Saaservisp 116
St. Germain 47,102,103
St. Léonard 27,98
St. Léonin, Bisse de 21,27
St. Luc 66
St. Marguerite, Bisse de 28
St. Marguerite, Chapel of 35,49,52,103

St. Maurice, Abbot of 28, 72
St. Théodule 25
Salgesch 13,27,71
Saltina 19
Samarin, Le 46,93,104, 105
Sand-trap, Sandfang 33, 57
Sanetsch, Barrage de, Col de 52
Sapin, Paroi du 35
Saracens 26
Sarreyer 72
Saut de la Matta 34
Savièse, Bisse de 8,12,21, 27,28,30,35,*35,36,37*, 44,46 et seq., *50,53,54,* 60,78,82,85,97,98,102
Savoy, Savoyards 25,26
Saxon, Bisse de 21,29,33, 60,65,68,71,82,100
Saxonna 58,79
Schattengspon 76
Schistose rock 16,19,91
Schnyder, Theo 7,81
Schöpfi 7,33,*34,4*
Schweizerische Landesbibliothek 8
Sealant 45,47,49,91
Sedimentary rocks 19
Seduni 25
Seepage 30,44,90,97
Seebach 109
Sérin 22
Service de l'Aménagement du Territoire de l'État du Valais 81,83
Sierre 71,106.107
Sillonin, Bisse de 27
Simpelberg 70
Simplon 16
Sion 11,16,20,25,27,29, 59,60,69,71,85,97
Sion, Bisse de 11,21,30, 45,54,*90*,92,93,103
Sion, Mayens de 54
Sionne 21,30,46,58,59, 105
Sixt des Eaux Froides 45
Soie, Château de la 27
Solution, chemicals in 22, 33
Sprinkler 64,76
Stalden 35,108,*113*
Staldenried 76
State of the VII dizains 25
Steffenbach gorge 52
Steinenbach 30

Stevenson, Douglas 85
Stigwassersuon *45*
Strahlen *87*
Subsidies 19
Suon 7
Suonengenossenschaft 59
Superimposition of bisses 44,54,*112-115*
Swiss Confederation 26,31
Siphon de Montorge 29
Siphon at Stalden *113*

Tablat 77
Tailla(z), Bisse 21,27,59, 71
Tassoz 60,61
Tavelli, Bishop 27
Tessel, tessle, tässlu 75
Thyon 2000 101
Tools 79
Törbel 76,108
Torin Nou I,28
Tornieu, torniou 56,79
Torrent Croix 52,*84*,85, 105
Torrent de Forniri *48*
Torrent, Neuf 28,82,103
Tour 78
Tourbillon 25
Tourniquet avertisseur 55,*56*
Trajou 28
Tranchant 56,*58*,79
Travelletti, Jean 8
Tretschbord 37
Tributaries of the Rhone 21
Trient 22,99
Tsâche, Torrent de la 101
Tsampi, Bisse de 52
Tsandra, Bisse de la 21,27, 28,52,72,95 et seq, 96,101
Tséna, tsénei 39
Turtmann,Turtmänna Torrent 16,22,107
Turtmannwerk 79

Undrasuon (Ausserberg) 78,*112*
Unterbach *113*

Valais, map of *4*
Valaisan Alps 15,16
Valère 25
Valpelline 21
Varen, Bisse de 21,27,54, 71,107
Vauclusian Springs 28,45
Vaud 19

Venetz 95,96
Vens 92
Verbier 100
Vercorin 27,72,*114*
Vex 71,
Vex, Bisse de 21,101
Vines, Vineyard 13,21,*29,* *30,*64,76,77,78,92,95, 97,106
Visp,Vispertal 27,107,108
Visperterminen 21,54, 66,89
Vissoye 78
Vogt 39
Vollèges 92,93
Vouasseur 51

Walliserdütsch 8
Wannigletscher 27,70
Wässerbeil 56,*58*
Wasserbüchlein 65
Wasserfassung 33
Wasserfuhr 7
Wasserhammer,(Water-hammer) 55,*56*,65,*6*
Wässerhaue 56,*58*
Wasserhüter 39,88
Wasserkehr 78
Wasserknebel 75
Wasserleite 7
Wässerplatte 56,*58*
Wasserscheit 75
Wasserschlegel 55,*56*,65,*6*
Wassertessel, Wassertessle 75
Wasservogt 39,89
Wastage 46,95
Water-rights 60,64,66,67, 68,75,95,96,97,98
Wildhorn 22
Wildstrubel 15
Wissbergschiene 76
Wysbach 70

Yvouettes, Les (Dorbon) 56

Zariré 66,67,68
Zeneggen 21,75,*113*
Zermatt 20,21
Ze Steinu 107
Zeuzier, Barrage de, Lac de 92,93,98,103,105
Zittoret, Bisse de 71

144